Structural Design
Against Deflection

Structural Design Against Deflection

Tianjian Ji

CRC Press
Taylor & Francis Group
Boca Raton London New York

CRC Press is an imprint of the
Taylor & Francis Group, an **informa** business

CRC Press
Taylor & Francis Group
6000 Broken Sound Parkway NW, Suite 300
Boca Raton, FL 33487-2742

CRC Press is an imprint of Taylor & Francis Group, an Informa business

No claim to original U.S. Government works

Printed and bound by CPI Group (UK) Ltd, Croydon, CR0 4YY on acid-free paper

International Standard Book Number-13: 978-1-138-61098-9 (Hardback)

Library of Congress Cataloging-in-Publication Data

Names: Ji, Tianjian, author.
Title: Structural design against deflection / by Tianjian Ji.
Description: Boca Raton, FL : CRC Press/Taylor & Francis Group, [2020]
Identifiers: LCCN 2019049039 (print) | LCCN 2019049040 (ebook) | ISBN 9781138610989 (hardback ; acid-free paper) | ISBN 9780367897932 (paperback) | ISBN 9780429465314 (ebook)
Subjects: LCSH: Structural design. | Deformations (Mechanics)—Mathematical models. | Tensegrity (Engineering)—Mathematical models. | Rigidity (Geometry) | Structural engineering—Case studies.
Classification: LCC TA658 .J523 2020 (print) | LCC TA658 (ebook) | DDC 624.1/771—dc23
LC record available at https://lccn.loc.gov/2019049039
LC ebook record available at https://lccn.loc.gov/2019049040

Visit the Taylor & Francis Web site at
www.taylorandfrancis.com

and the CRC Press Web site at
www.crcpress.com

Contents

Preface

Deflection is a major consideration in the design of structures. Buildings have become taller, bridges longer and floors wider, and these all involve the control of the structural deflection. Designing structures against deflection is not only concerned with serviceability and stiffness, but is also related to safety, such as failure due to buckling or excessive stress, efficiency and even the elegance of structures. Therefore, it is necessary to take a holistic approach to examine this topic.

Different claims are often made about form, deflection and internal forces in structures. It is normally thought that *structural form determines deflection and internal forces*. It has also been said that *structural form is determined by the flow of internal forces*. The two statements appear contradictory, but they have one thing in common in that form, deflection and internal forces are so closely related that altering any one will change the other two. Choosing the form of a structure can also be seen as selecting the path and distribution of internal forces in the structure which extends the way to producing efficient designs.

Many well-known engineers believe that the appropriate use of structural principles will lead to efficient and elegant structures. For example, Professor David Billington at Princeton University wrote "the best engineers followed certain general principles of design to arrive at fine works, and these general principles allowed for their own specific and personal vision of structures" in his book, *The Tower and the Bridge*. Professor Mike Schlaich, partner of SBP, wrote "Elegance appears when the challenging task of fusing the principles of good structures seems to be achieved without much effort." These statements pose three questions: What are the general principles or the principles of good structures? How can such principles be creatively implemented into the design of structures? And how can the principles and the routes to implementation be explicitly expressed and passed on to other engineers, especially the engineers of the next generation. This book tries to answer the three questions by identifying some of the principles and concepts relating to deflections and internal forces, demonstrating their effectiveness and efficiency and examining their creative implementation in existing structures.

In parallel with the previous thoughts, the idea to write this book originated from my teaching courses on structural engineering to final-year undergraduate and postgraduate students at the University of Manchester. New teaching contents have been developed to help students' understanding from structural elements to whole structures and from theory to practical application, which form the basis of the book and are organized through the common thread of deflection and internal forces.

The presentation of the book also follows five philosophical criteria:

1. *Seeking new connections between theory and practice.* It has been said that there is a gap between theory and practice. How can this gap be bridged? When crossing a wide river, a bridge may require several intermediate supports. Similarly, new intermediate connections need to be sought between theory and practice, such as those which exist between theory and structural concepts, between the structural concepts and physical measures to implement them and between the implementation measures and practical cases. Connections have also been sought between examples with and without involving implementation measures that are developed based on one of the structural concepts. This allows illustrating and quantifying the effectiveness and efficiency of the implementation measures and the corresponding structural concepts. Practical cases have also been connected with simplified hand calculation models to reveal the effect of the embedded structural concepts.

2. *Exploring new meanings of structural theory.* It is thought that structural theory is a mature subject. However, it is still possible to explore new meanings from old theories. New meanings of the virtual work principle are explored and interpreted leading to a set of four structural concepts. The structural concepts reveal the relationships between deflection and internal forces of truss and frame structures. They form the basis of this book showing that smaller deflection can be achieved through generating more desirable distribution of internal forces in a structure. It is noted that more desirable distribution of internal forces can also lead to more effective, efficient and elegant structures.

3. *Being simple.* It is a common belief that a basic and simple theory often has wide application, *i.e.* simple and universal, such as the Newton's second law. What is a simple and universal theory for structural design? This question will be examined in Chapter 2 of this book. Another way is to make the presentation of theory simple allowing many structural engineers to use it. Four structural concepts, abstracted from basic theory, will be presented as "rules of thumb" for easy understanding and for practical use. It is also believed that a problem, an equation or a structural phenomenon can be explained in a simple manner while its physical essence is captured. This way of explanation is termed as intuitive interpretation in this book, which is an effective tool and skill and will be demonstrated using examples.

4. *Evolving into intuitive understanding.* When the understanding of theory evolves into an intuitive understanding, it will help to lead to appropriate and good use of theory. Structural design, including the design against

deflection, does not start from theory. Instead, it starts from the intuitive understanding of structural behaviour and structural adequacy. For developing such intuitive understanding, a number of hand calculation examples, which are abstracted or simplified from practical cases, are studied quantitatively and comparatively between with and without involving one of the four structural concepts.

5. *Making wide and wise applications of theory.* A number of practical cases, linking with the hand calculation examples, demonstrate that the four structural concepts presented have been used widely and have provided clever solutions to challenging engineering problems. The routes to implementation of the structural concepts into the design of structures are explicitly listed and discussed to promote wider and wiser use in practice. It is hoped that the reader will be stimulated by the examples and cases presented to make their own creative applications.

This book contains material on three structural concepts relating to stiffness developed in an earlier book, *Understanding and Using Structural Concepts*, by Tianjian Ji, Adrian Bell and Brian Ellis, to which this book is a successor. Anyone who has used *Understanding and Using Structural Concepts* will find the present book substantially more comprehensive on the understanding and on the application of the three structural concepts.

Structural design against deflection requires a broad knowledge of materials, analysis, structural behaviour, loading, environment, construction details, etc. This book focuses on structural concepts and their implementation in practice to achieve more effective, efficient and perhaps even more elegant structures.

This book integrates teaching, practice and research through the common thread of structural concepts for structural design against deflection. It is hoped that the book provides an inspirational experience to advanced undergraduate and graduate students studying civil engineering and architecture, and enhances the holistic comprehension of structural engineers and architects.

Tianjian Ji
The University of Manchester, UK

Acknowledgements

I express my sincerest gratitude to Dr. Adrian Bell, my colleague and co-author of *Understanding and Using Structural Concepts*, for his help in checking the manuscript of this book, for his constructive comments and for making the book more readable. I am also very grateful to Dr. Brian Ellis, my previous colleague and also co-author of the previously mentioned book, who was able to read and check the first two chapters of this book on a short notice.

The writing of this book has been constrained by my available photographs and figures of actual structures. Several individuals have however helped me in this respect and their contributions are acknowledged next to relevant photographs in the text. In particular, I am indebted to Mr. Nicolas Janberg, owner and creator of structurae.net, Germany, for providing several photos with high quality.

I am very grateful to Mr. Andrea Codolini, a Ph.D. student at the University of Manchester, for producing some of the drawings in this book.

The assistance of Taylor & Francis in the publication of this book is greatly appreciated. I would like to thank Mr. Tony Moore, Senior Editor, for his encouragement in writing this book. Gabriella Williams and Lisa Wilford, at the UK office, and Jennifer Stair, Production Editor, at the USA office, provided assistance at different stages. I am also grateful for the help from Denise File and her team at Apex CoVantage in preparing the final version of the manuscript.

Author Bio

 **Dr. Tianjian Ji** is Reader in Structural Engineering at the University of Manchester, a Chartered Engineer and Fellow of the Institution of Structural Engineers and Higher Education Academy, UK. He graduated in Civil Engineering and received MSc degree in Structural Mechanics both from Harbin Institute of Technology, China, and PhD degree in Civil Engineering at the University of Birmingham, UK. Before joining Manchester in 1996, he had worked in industry for over ten years, mainly at Building Research Establishment, UK, and China Academy of Building Research, Beijing.

His research area is in vibration and structural dynamics using both experimental and theoretical methods, including structural vibration induced by human rhythmic crowd loads, human-structure interaction, probabilistic seismic risk assessment of nuclear power plants and finite element modelling of structural dynamic behaviour. He has supervised 14 PhD students and over 70 MSc students, and published over 140 articles.

He has taught at all year levels on Mechanics, Structural Analysis, Structural Design, Research Methods and Earthquake Engineering. He led the development of what is called *Seeing and Touching Structural Concepts* for helping students to develop an intuitive understanding of structural concepts (www.structural-concepts.org). A book was published with the same title in 2008 and the second edition of the book with a revised title, *Understanding and Using Structural Concepts*, was published in 2016. Both editions were translated into Chinese and published in China. He received the *Award for Excellence in Structural Engineering Education* from the Institution of Structural Engineers in 2014 and the *Teaching Excellence Award* from the University of Manchester in 2016.

He has actively taken consultancy work for helping to solve practical vibration problems, including some most prestigious structures in the UK, such as vibration measurement of the London Eye, and for providing or reviewing remedial schemes of building floors subjected to rhythmic crowd loads or impact loads.

Chapter I

Introduction

1.1 Deflection of Structures

For the structural design of a building, engineers need to check deflection, vibration, stability and strength of the structure and its components, and ensure that they satisfy all requirements, *i.e.* they have appropriate values smaller or larger than limiting values. Deflection and vibration are classified as serviceability problems while stability and strength are considered to be safety problems. These four issues are normally analysed and checked independently; but are there any connections between the four of them?

The deflection of structures is a key serviceability consideration and may often control the design of slender floors, tall buildings and long bridges. As buildings become taller, bridges longer and floors wider, the associated deflections of these structures become major design issues.

Deflection limits are applied to structural elements, such as beams and floors, and to whole structures, such as buildings and bridges. The limits often require that the possible maximum deflection of a structure or a structural element should be smaller than a certain value. For example, the limit for the maximum deflection of a truss structure is 1/180 of its span [1.1]. For a defined structure and a given loading, the deflection of the structure is calculated using the following equilibrium equation:

$$[K]\{U\} = \{P\} \tag{1.1}$$

where $[K]$ is the stiffness matrix that is related to the structural form and the cross-sectional and material properties of the structural members, $\{U\}$ is the deflection vector to be determined and $\{P\}$ is the given loading.

Structural vibration is another type of serviceability issue, which may cause discomfort to users of the structure and restrict the functionality of the structure. Structural vibration is not only related to the dynamic loads applied but is also related to the dynamic properties of the structure, *i.e.* natural frequency, damping ratio and modal mass or modal stiffness. In the design of grandstands and floors used for rhythmic activities, one design philosophy requests that the fundamental natural frequency of the structure should be larger than a certain value to avoid possible resonance [1.2, 1.3]. The natural frequencies and the mode shapes of the structure can be determined by solving the following eigenvalue equations:

$$([K] - \omega^2 [M]) \{\phi_v\} = \{0\} \tag{1.2}$$

where $[M]$ is the mass matrix, ω is the circular natural frequency and $\{\phi_v\}$ is the vibration mode of the structure. The stiffness matrix $[K]$ is the same as that in equation 1.1.

Stability of a structure or a structural member is considered as a safety problem. When a structure is subjected to external loads and self-weight, compressive forces/stresses are induced in the body of the structure. In such a situation, engineers need to check if the whole structure will lose its stability and if any individual member will buckle. Quite often the buckling of a compression member can result in a sudden failure of the member which may lead to the development of a mechanism and local or even global failure of the structure. The global stability of a structure is evaluated by a similar eigenvalue equation to that for natural frequency:

$$([K] - \lambda [K_G]) \{\phi_s\} = \{0\} \tag{1.3}$$

where $[K_G]$ is the geometric or initial stress stiffness matrix that is formed based on the applied loads and the structural form, λ is the buckling load factor (λ times the existing loads would cause global instability of the structure) and $\{\phi_s\}$ is the bucking mode of the structure which describes the pattern of instability.

Strength measures the capacity of individual structural members to withstand the internal forces applied to them by the external loads on the structure. Unlike deflection, vibration and stability, strength is considered for individual members rather than for the whole structure, but the failure of an individual member may lead to an unsafe structure. Once the internal force in a member is determined, the corresponding stress is easily calculated and compared with its allowable stress. If the stress is larger than the allowable stress, the cross-section of the member may need to be enlarged.

The relationship between deflection and bending moment of a uniform beam is:

$$EI \frac{d^2 u(x)}{dx^2} = -M(x) \tag{1.4}$$

where $u(x)$ and $M(x)$ are the deflection and bending moment at coordinate x of the beam, and EI is the rigidity of the cross-section of the beam.

For a plane element in finite element analysis, the relationship between strain $\{\varepsilon\}$ and nodal displacement $\{d\}$ is defined as:

$$\{\sigma\} = [E]\{\varepsilon\} = [E][B]\{d\} \tag{1.5}$$

Table 1.1 Relationships between structural design problems

Type of problem	Strength	Deflection	Free vibration	Stability
Basic equation	$EI\dfrac{d^2u(x)}{dx^2} = -M(x)$ $\{\sigma\}=[E][B]\{d\}$	$[K]\{U\} = \{P\}$	$([K]-\omega^2[M])\{\phi_v\}$ $= \{0\}$	$([K]-\lambda[K_G])\{\phi_s\}$ $= \{0\}$
Relation to deflection	The internal force and stress are directly related to deflection.	Deflection, natural frequency and buckling load factor are all related to stiffness (the stiffness matrix).		

where $[B]$ is the strain-displacement matrix that transfers the nodal deflections of the element to the strains within the element, and $[E]$ is the material property matrix. The nodal displacement $\{d\}$ of the element is taken from the global displacement $\{U\}$ in equation 1.1.

It can be observed from equations 1.1–1.5 that:

- For deflection, vibration and stability problems, equations 1.1–1.3 contain the stiffness matrix $[K]$ of the structure and show qualitatively that *the stiffer the structure, the smaller the deflection, the higher the natural frequency and the larger the buckling load factor.*
- As the deflection vector and stiffness matrix are "reciprocal" of each other for a unit load vector, the previous statement can be rewritten as: *the smaller the deflection of a structure, the higher the natural frequency and the larger the buckling load factor.*
- For a strength problem, equations 1.4–1.5 show that the *internal forces or internal stresses are directly related to deflection.*

The relationships between the four structural design problems are summarised in Table 1.1.

It can be seen from Table 1.1 that deflection is a physical quantity that is directly related to internal forces or stresses (equations 1.4–1.5) and is indirectly related to the natural frequency and the buckling load factor (equations 1.1–1.3).

For explicitly expressing the relationships between deflection and natural frequency, between deflection and buckling load, and between deflection and internal forces, consider a simply supported uniform beam with a length of L, cross-sectional rigidity of EI and a uniformly distributed mass of m.

a) Deflection

The maximum deflections of the beam due to its self-weight, mg, and a concentrated load, F, at its centre are respectively:

$$\Delta_q = \frac{5mgL^4}{384EI} \quad \text{and} \quad \Delta_F = \frac{FL^3}{48EI} \tag{1.6, 1.7}$$

where g is the acceleration due to gravity.

b) Fundamental Natural Frequency and Deflection

The fundamental natural frequency of the uniform beam is:

$$f = \frac{\pi}{2}\sqrt{\frac{EI}{mL^4}} \qquad (1.8)$$

It can be seen that equations 1.6 and 1.8 both contain mL^4/EI that gives the connection between the fundamental natural frequency and the maximum deflection. Eliminating mL^4/EI in the two equations gives the relationship between the fundamental natural frequency and the maximum deflection:

$$f = \frac{17.75}{\sqrt{\Delta_q}} \qquad (1.9)$$

In this calculation, g is taken as 9810 mm/s^2 and Δ_q is in mm. Equation 1.9 shows that *the fundamental natural frequency of a beam is inversely proportional to the square root of the deflection*. In general, *the smaller the deflection, the larger the fundamental natural frequency*. Equation 1.9 has been used in several design guides [1.4] to facilitate a quick estimation of the fundamental natural frequency without conducting an eigenvalue analysis.

c) Critical (or Buckling) Load and Deflection

When the beam is subjected to a compressive load P applied at its ends along its longitudinal axis, the critical load is:

$$P_{CR} = \frac{\pi^2 EI}{L^2} \qquad (1.10)$$

Substituting equation 1.7 into equation 1.10, by removing EI, leads to:

$$P_{CR} = \frac{FL\pi^2}{48\Delta_F} \qquad (1.11)$$

Equation 1.11 indicates that *the buckling load of a strut is inversely proportional to the lateral deflection of an equivalent beam caused by a concentrated load acting at its centre*. Equation 1.11 also suggests that the buckling load of a strut can be experimentally determined by conducting a non-destructive bending test [1.5].

d) Bending Moment and Deflection

The maximum bending moment in the beam due to its self-weight is:

$$M_q = \frac{mgL^2}{8} \qquad (1.12)$$

The relationship between the maximum bending moment M_q and the maximum deflection Δ_q can be derived from equations 1.6 and 1.12 as follows:

$$M_q = \sqrt{1.2mgEI\Delta_q} = \frac{48EI}{5L^2}\Delta_q \qquad (1.13)$$

Equation 1.13 shows qualitatively, at a structural element level, that *the smaller the maximum deflection, the smaller the maximum bending moment.*

Seeking the connections between deflection, natural frequency, buckling load and internal force not only helps gain a better understanding but also leads to wider and wiser applications, such as estimating the fundamental natural frequency using a known deflection and determining the buckling load by conducting a bending test.

A question then arises how to better design structures against deflection by reducing deflections which also helps to increase the fundamental natural frequency and buckling load capacity of the structures and reduce internal forces. Therefore, there is a need to return to basics and examine the relationships between deflection, structural form and internal forces in addition to applied loading.

1.2 Form, Deflection and Internal Forces

It is often thought that *structural form determines internal forces in a structure.* This understanding can be based on an input-structure-output model as follows:

Figure 1.1 Relationships between input, structure and output.

where loading is the external forces applied on a structure; structural form describes the global structural system that also embraces architectural form, and internal forces are the forces in structural members resulting from the loading on the structure and the structural form, which normally include axial forces, shear forces and bending moments. When a structural form is designed or selected and the structure is subjected to a given load (input), the deflections and internal forces (output) can then be uniquely determined (Figure 1.1), *i.e.* the output is the consequence of the response of the structural form to the given set of loads, *i.e.*:

$$[K]\{U\} = \{P\} \text{ or } \{U\} = [K]^{-1}\{P\} \qquad (1.1)$$

The internal forces in members of the structure are then normally determined based on the calculated deflections. These deflections and internal forces may then be used as feedback to revise the geometry of the structure and the dimensions of its members, which leads to a change of the stiffness matrix in equation 1.1, to achieve an improved design.

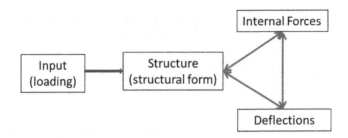

Figure 1.2 Relationship between form, internal forces and deflections of a structure.

Alternatively, it may be said that *the form of a structure is determined by the flow of internal forces*. In fact, structural form, internal forces and deflections of a structure subjected to a given loading are closely related and interact with each other. In order to reveal and examine the relationships between structural form, internal forces and deflections, Figure 1.1 may be revised as shown in Figure 1.2. This indicates that altering internal forces can change deflections and structural form and controlling deflection can also revise structural form and internal forces, in addition to varying structural form can lead to a new set of internal forces and deflections. For example, if the deflection at a given point in a particular direction is constrained, this will correspond to a change of the structural form by requiring an appropriate support at the point and this alters the stiffness matrix. Altering internal force paths and varying structural form occur simultaneously although the magnitudes of the internal forces are determined after the structural form has been confirmed.

It is unlikely that the relationships between structural form, internal forces and deflections of a structure can be expressed explicitly. However, it is possible to gain a qualitative understanding of them through examining two similar plane frames.

Question

Figure 1.3 shows two four-bay and four-storey plane pin-jointed structures with the same dimensions. All the members are made of the same material (E) and have the same cross-sectional area (A). The vertical and horizontal members have the same length (L). The same concentrated loads of 0.5N are applied anti-symmetrically at the two top corners in the horizontal direction on each frame. Two bracing members are required to be placed in each storey, hence the two frames use the same number of bracing members, *i.e.* the same amount of material. The only difference between the two frames is the arrangements of bracing members, which can be discussed as follows:

Frame A: Four bracing members are placed symmetrically on each of the two side bays and are arranged in the same orientation, *i.e.* the bracing members are not directly linked. As all eight bracing members are placed on the side bays, the two middle bays have no bracing members. This type of bracing pattern is often seen in practice.

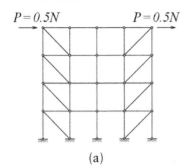
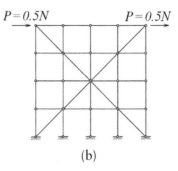

(a)   (b)

Figure 1.3 Two plane pin-jointed frames with different bracing arrangements. (a) Frame A: the bracing members are arranged in parallel in the two side bays. (b) Frame B: the bracing members are arranged across the four bays.

Frame B: The bracing members are arranged across the four bays of the frame and are linked in straight lines. This bracing pattern can be generated from that in Frame A by: 1) changing the orientation of the two bracing members on the ground (first) level; 2) moving the two bracing members in the second level horizontally inward to the next panels and altering their orientation; and 3) moving the two bracing members in the third level horizontally inward to the next panels.

With the structure form and the loading defined, the internal forces in the members and the maximum deflections of the two structures can be determined and the relationships between the form, deflections and internal forces for these particular structures can then be examined.

Solution

The two structures are statically indeterminate. However, they are both symmetric structures subjected to anti-symmetric loads. According to the structural concept that *a symmetric structure subjected to anti-symmetric loading will result in only anti-symmetric responses (internal forces and deflections),* the four central vertical members will have to be in a zero-force state and the nodal points along the central vertical members of the two frames will have no vertical displacements. Therefore, the two frames can be simplified and equivalently represented by their left halves with appropriate boundary conditions as shown in Figure 1.4. Each half frame has 16 vertical and horizontal members and four bracing members.

It can be noted that the middle vertical members are removed as there are no internal forces in these members and the vertical displacements of the points along the middle members are constrained using roller supports. Now the two half frames become statically determinate structures and the internal forces of all the members can be directly and easily calculated by hand.

The internal forces in the members of the two simplified frames can be determined using the equilibrium conditions at the pinned joints and the calculated internal forces in the members of the two half frames are as shown in Figures 1.4 (a) and (b), where positive values indicate the members in tension and negative values indicate the members in compression. In addition, the internal force paths (non-zero force members) are indicated using dashed lines.

The maximum lateral deflections at the loading positions of the two half-frames can be determined using a well-known equation, taken from textbooks [1.6, 1.7], as:

$$\Delta_{A,\max} = \sum_{i=1}^{20} \frac{N_i^2 L_i}{EA}$$

$$= \frac{L}{EA}\left[\left(\frac{1}{2}\right)^2 \times 5 + (1)^2 \times 2 + \left(\frac{3}{2}\right)^2 \times 2 + (2)^2 + \left(\frac{\sqrt{2}}{2}\right)^2 \sqrt{2} \times 4\right] \times 2$$

$$= \frac{11.75 + 2\sqrt{2}}{EA} \times 2 = \frac{29.16L}{EA}$$

$$\Delta_{B,\max} = \sum_{i=1}^{20} \frac{N_i^2 L_i}{EA} = \frac{L}{EA}\left[\left(\frac{1}{2}\right)^2 \times 4 + \left(\frac{\sqrt{2}}{2}\right)^2 \sqrt{2} \times 4\right] \times 2$$

$$= \frac{\left(1 + 2\sqrt{2}\right)L}{EA} \times 2 = \frac{7.656L}{EA}$$

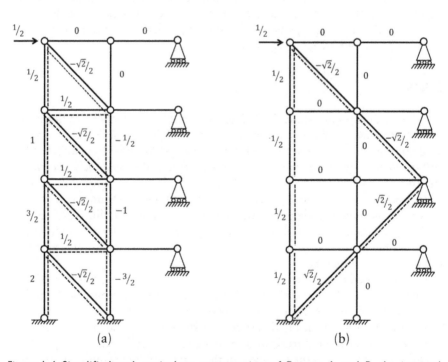

Figure 1.4 Simplified and equivalent presentation of Frames A and B, the internal force paths in dashed lines and values of the internal forces. (a) Frame A. (b) Frame B.

The ratio of the two deflections is:

$$\frac{\Delta_{B,max}}{\Delta_{A,max}} = \frac{7.656L}{EA} \times \frac{EA}{29.16L} = 0.263$$

The maximum lateral deflection of Frame B is only 26.3% of that of Frame A although the same number of members is used in both frames. This demonstrates the significant effect of structural form on the deflection and the internal forces in members as the two frames use the same amount of material. The paths and values of the internal forces in the members of the two frames and the characteristics of the bracing patterns can be observed and examined to provide an intuitive understanding of the reasons why Frame B has much smaller deflections than Frame A:

1. There are more zero-force members in Frame B than in Frame A. There are 12 zero-force members in Frame B compared with six in Frame A.
2. The magnitudes of the internal forces in the members in Frame B are smaller than those in Frame A. The largest absolute internal force in a member is 0.5 in Frame B compared with 2.0 in a member in Frame A without considering the constant internal forces in the bracing members, which are the same in the two frames.
3. The differences between the magnitudes of the internal forces in the members of Frame B are smaller than those of Frame A. The largest absolute difference between the internal forces in members is 0.207 in Frame B compared with 1.5 in Frame A, ignoring the members with zero-force.
4. The four-bay, four-storey Frame B is braced globally while Frame A is braced locally in its two side bays.
5. It may be considered that Frame B looks more pleasing and elegant than Frame A (Figure 1.3)

The first three observations show that the characteristics of the internal force paths and distributions in the two frames are clearly technical issues. The fourth observation is about the geometry or pattern of the bracing members, which is a design issue. The fifth observation concerns the appearance of the two frames, which is related to a human perception of the quality and beauty of a structure. It appears that *more zero-force members, smaller internal forces in members and a more uniform distribution of internal forces in members lead to smaller deflections*. These observations from the two frame examples are interrelated and inspire the thought that the internal force flow and its distribution can be positively designed to define the structural geometry and topology, and to control deflections. The observations generate the following three questions:

1. What are the rules or structural concepts embedded in Frame B, which result in Frame B having much smaller deflections than Frame A without

using more material? Are such rules or structural concepts applicable to the design of other structures?

2. How can the flow and distribution of internal forces be actively considered to aid the design of structural form?
3. How can internal forces be designed to make a structure more effective (smaller deflections), more efficient (less materials) and perhaps more elegant?

Answering these three questions requires a harmonious combination of intuitive understanding of structures and a sound technical knowledge of structures, which will be discussed in Chapter 2.

1.3 Intuition of Structures

According to Mario Salvadori (a structural engineer and professor of both civil engineering and architecture at Columbia University), who wrote a forward for Torroja's book [1.8], outstanding engineers, like Eduardo Torroja (Spanish structural engineer and architect), reached very high levels through four phases: 1) devoting their early years to a long and thorough study of fundamentals; 2) applying the fundamentals to the solutions of original problems in practice and accumulating experience; 3) slowly synthesising their accumulated experience to reach what is called "intuition"; and 4) bringing them to higher and higher levels with ever-decreasing effort and ever-increasing enjoyment of their work.

Pier Luigi Nervi (Italian structural engineer and architect) said that the mastering of structural knowledge is the result of a physical understanding of the complex behaviour of a building, coupled with an intuitive interpretation of theoretical calculation.

These thoughts from eminent engineers indicate the importance of intuitive interpretation of theoretical calculation and structural behaviour and the ways of developing intuition. They have also led to thinking about what intuitive interpretation means and how intuition could be learned at an earlier stage or taught at university [1.9].

Intuitive knowledge, intuitive understanding and intuitive interpretation are related but they have different meanings and characteristics.

1.3.1 Intuitive Knowledge

Such knowledge often comes from experience which is correct but may not have theoretical support or the theory behind the knowledge is not available or is not known. For example, many families know that rubber footpads reduce the vibration generated by washing machines. However, most do not know the reason why the small pads can effectively reduce the vibration, but they can still make a good use of the knowledge. This type of knowledge can be gained from personal experience or learned from the experience of others.

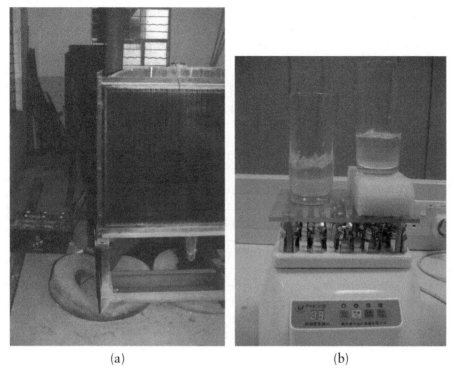

(a)   (b)

Figure 1.5 Vibration isolation. (a) Tyres used for isolation in practice.(Courtesy of Professor Biaozhong Zhuang, Zhejiang University, China). (b) Demonstration of vibration isolation in teaching.

Figure 1.5 illustrates two isolation measures used for vibration reduction. Figure 1.5a shows tyres placed between the ground and a generator in a rural area of a developing country. The presence of the tyres led to a lower natural frequency of the generator-tyre system and moving it away from the operating frequency of the generator. The operators of the generator had not received a university education and were not aware of vibration theory, but they knew from their own experience, or the experience of others, that the presence of the tyres could reduce vibration. Figure 1.1(b) shows a laboratory demonstration of the effect of vibration isolation to students at a university. A medical shaker was used as a shaking table to generate harmonic base movements in three perpendicular directions. One glass is fixed directly to the table of the shaker and a similar glass is glued to a layer of plastic foam that is fixed to the table. The two glasses are filled with similar amounts of water. When the shaker moves at a preset frequency, it can be observed that the water in the glass mounted on the plastic foam moves less significantly than that in the other glass. The difference in movements is due to the effect of the plastic foam that isolates the base motion of the glass above. The plastic foam and the glass and water above form a new system that has a much lower natural frequency than that of the glass with water alone. This demonstration has been shown to students to enhance their understanding of vibration reduction and to audiences of the general public to help them gain the intuitive knowledge that isolation can reduce vibration.

The two examples of vibration isolation indicate that theory can be illustrated and practical cases can be simulated using physical models to produce a broader perspective and gain intuitive knowledge.

1.3.2 Intuitive Understanding

Such understanding of a problem can be gained from observations and from practical experience or/and from fundamental theories. It often comes without conscious learning or theoretical derivations. It is observed that a person who has many years of practical experience and a sound theoretical foundation is able to gain an intuitive understanding of a problem.

Figure 1.6 shows the North Stand at Twickenham, UK, in which vibration measurements were taken on the middle cantilever tier when the stand was empty and when the stand was full of spectators. Figure 1.7 shows the measured response spectra of the tier, without spectators and with spectators. Comparing the two spectra, three significant phenomena were apparent [1.10]:

Figure 1.6 The North Stand, Twickenham.

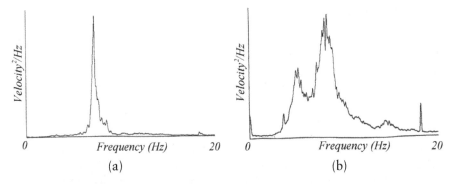

Figure 1.7 Response spectra of the North Stand, Twickenham. (a) Without spectators. (b) With spectators.

1. An additional natural frequency was observed in the occupied stand.
2. The damping increased significantly when spectators were present.
3. The natural frequency of the empty stand was between the two natural
   frequencies of the occupied stand.

The observations were contrary to the belief that a human body acts as an inert
mass in structural vibration [1.11]. If the spectators acted as inert masses, the
occupied stand would have only one natural frequency that should be smaller
than that of the empty stand, and the inert body masses would not increase the
damping of the occupied stand. The intuitive understanding of the experimental
observations was that the spectators did not act as inert masses on the stand in
the vertical structural vibration. This intuitive understanding was an outcome
of the site observations and some knowledge of fundamental vibration theory
and has led to much research on the new topic of human-structure dynamic
interaction [1.10].

Students at universities learn structural theory but they may not often have
opportunities to observe structural behaviour and conduct experiments. However, it is possible to produce physical models and to show related practical
examples for students to appreciate.

1.3.3 Intuitive Interpretation

Intuitive interpretation means that an equation, an observation or structural
behaviour can be explained in a simple manner, while the explanation captures the physical essence of the problem. This often results from a sound
understanding of theoretical fundamentals and from practical experience.
Intuitive interpretation in structural engineering is an effective tool to explore
new meanings, seek new connections, develop new understanding and promote wide and wise applications. It is best to illustrate intuitive interpretation
using examples.

1.3.3.1 Mathematical Equations

Many equations can be used to practice intuitive interpretation and to gain an improved understanding of theory, leading to practical applications and appreciation of what intuitive interpretation means. For example, second moment of area of a plane cross-section is expressed as:

$$I = \int y^2 dA \tag{1.14}$$

where y is the distance between the neutral axis of the cross-section and dA that is the area of an infinitely small area. Second moment of area is the geometrical property of the section which is related to its area and to the distribution of the area. Students were asked to interpret equation 1.14. One of the answers was that *the second moment of area of a cross-section is the sum of the products of a small area and the square of the distance between the centre of the area and the neutral axis of the section.* This statement is correct but is actually a verbal expression of equation 1.14 rather than an intuitive interpretation that tends to capture the physical essence of the equation. The intuitive interpretation of equation 1.14 should be: *the further (closer) the material is away from (to) the neutral axis of a section, the larger (smaller) the contribution to the second moment of area of the section.* It is this interpretation, or understanding, that forms a basis for creatively designing the shape of a cross-section of a beam, such as I-sectioned beams or cellular beams. As tall buildings can be treated as cantilevers in conceptual designs, shear walls and columns should be arranged as far away as possible from the neutral axis of the building plane. Equation 1.14 provides a means to calculate the second moment of area of a cross-section while the intuitive interpretation of equation 1.14 paves a way for creative applications.

1.3.3.2 Observation of Structural Behaviour

Figure 1.8a shows a test rig, equipment and the specimen that were used in a vibration-buckling test. A straight steel strut was placed in the test rig with the two ends of the strut having pinned supports. Weights were added gradually to apply compression to the strut until it buckled. In parallel with the buckling test, the fundamental natural frequency of the loaded strut in the lateral direction was measured at each loading stage using a small accelerometer placed at the centre of the strut and linked to a vibration analyser. At each loading stage, a gentle lateral impact was applied to the strut (a tap from a finger) to generate lateral vibrations. The weights and the natural frequency at each loading stage were recorded. Figure 1.8b shows the relationship between the measured natural frequency squared, on the vertical axis, and the applied vertical load, on the horizontal axis. The points show the measurements and a straight line is fitted to the points.

Students were asked to interpret the observations and the results shown in Figure 1.8. One answer was that *there is a linear relationship between the natural frequency squared and the compressive force.* This is an obvious

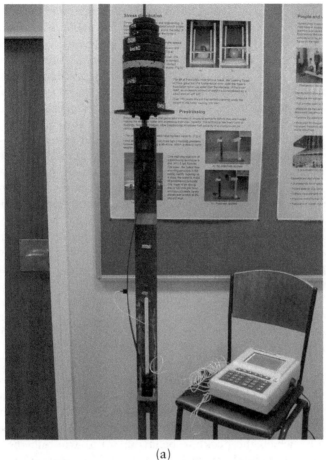

(a)

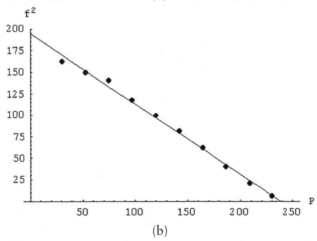

(b)

Figure 1.8 Vibration-buckling experiment. (a) Vibration-buckling test of a loaded strut. (b) Relationship between natural frequency squared and the compression force.

observation from Figure 1.8(b). However, there is a more important observation: *when the strut reaches its buckling load, its fundamental natural frequency becomes zero*, which corresponds to the intersection point of the fitted line and the horizontal axis in Figure 1.8b. As the lateral stiffness of the strut is proportional to its fundamental natural frequency squared, the intuitive interpretation of this observation is that: *a strut buckles when it loses its lateral stiffness*. This interpretation provides an alternative definition of buckling. When checking current textbooks, one can note that the current definition of buckling of a strut is a description of the phenomenon of buckling.

This experiment and observation have generated further discussion as to whether frequency measurements can provide a non-destructive method for predicting the buckling load of a real structure. Taking two frequency measurements at different loading magnitudes and drawing a straight line passing through the associated two points in Figure 1.8b, the intersection point of the line and the horizontal axis is the buckling load. This requires high quality measurement of the fundamental natural frequencies of the strut at two different loading states for application. Such a linear relationship between P and f^2 may not exist for other types of structures.

This example demonstrates how intuitive interpretation expresses an important observation of structural behaviour in a simple manner while capturing its physical essence of buckling, and can also be used to illustrate the philosophic criteria embedded in the presentation of the book:

Seeking new connections: Buckling of a strut and free vibration of a beam are two different problems in textbooks and in engineering design, and they are normally considered independently. The new connection between the two problems was established through the experimental set-up by which the buckling behaviour of a strut and the vibration behaviour of a simply supported beam could be examined simultaneously in some details.

Exploring new meanings: Following the new connection and the experiment, the new meaning of buckling was explored, interpreted and stated concisely as *a strut buckles when it loses its lateral stiffness* that is an alternative definition of buckling to complement with the existing one: *a strut buckles when bending occurs*.

Being simple: For conducting a combined buckling and vibration experiment, a strut may be the simplest structural member. The experiment used an existing rig for buckling tests of struts, which made the experiment simple and straightforward. A simple frame model could be used for conducting similar tests, but it would require more efforts both experimentally and theoretically.

Evolving into intuitive understanding: The fundamental natural frequency also indicates the remaining buckling capacity of a structure. When the fundamental natural frequencies of similar racking systems with different loads can be measured, these data would help estimate the remaining buckling capacity of the structures.

Making wide and wise application: This study encouraged further studies to examine the possibility and conditions for developing a non-destructive experimental method that natural frequency measurements are used to predict buckling loads of structures. Following the same route of this experiment, a new connection between a bending test and a buckling test was sought, which led to a conclusion that a bending test can be used to predict the buckling load of the test member [1.5].

1.3.3.3 Hand Calculation

Hand calculation is an effective skill which can facilitate intuitive interpretation. It may be a necessity for intuitive interpretation that one should be able to simplify a complex structure into a model that still retains the physical essence of the structure but is simple enough to be analysed by hand.

Different bracing patterns can be observed in existing structures, such as tall buildings, temporary grandstands and scaffoldings. Real structures are three dimensional and are too demanding for hand analysis. For hand analysis, there is a need to create simple structure models abstracted from the real structures which possess the physical essence of the bracing patterns. This has been demonstrated in Figure 1.3 in Section 1.2. The hand calculations for the two simple frames has provided the necessary results for comparison and for intuitive interpretation, which help to identify new meanings, new connections and new understanding of the relationships between form, internal forces and deflection.

1.3.3.4 Definition of Structural Concepts

In an earlier work [1.5], intuitive interpretation is used to define structural concepts as

> *A structural concept is an intuitive interpretation and concise representation of a mathematical relationship between physical quantities, which captures the essence of the relationship and provides a basis for practical applications in structural engineering.*

This definition clearly states that structural concepts come from the intuitive interpretation of mathematical equations. Such interpretation can be applied to observations, structural behaviour and results from hand and computer calculations. The intuitive interpretation of equation 1.14 and Figure 1.8 are examples in which two structural concepts are identified and presented concisely.

The illustrations in this subsection indicate that using models, practical examples, observations and calculation results can create scenarios for effectively helping students to gain intuitive knowledge and intuitive understanding and to practice intuitive interpretation, which complements the contents of textbooks.

1.4 Design against Deflection Based on Beam Theory

Structural design against deflection requires the use of equations to calculate the deflections. There is a simple equation for the central deflection of a uniform beam in textbooks [1.6]:

$$\Delta_{max} = \alpha \frac{qL^4}{EI} \tag{1.15}$$

where Δ_{max} is the maximum deflection, q is a uniformly distributed load, L is the span, E modulus of elasticity and I the second moment of area of the cross-section of a beam. α is a non-dimensional coefficient relating to boundary conditions, for example, 5/384 for a simply supported beam and 1/8 for a cantilever. This equation is explicit and clearly shows the relationship between deflection and five other parameters. Implementation of equation 1.15 for reducing deflection has been used in practice [1.12] as outlined following a "rule of thumb" format:

1. *Reducing span L:* As the deflection is proportional to L to the power of four, reducing span where possible is the most effective way to reduce deflection, *e.g.* via the provision of additional supports. Figure 1.9 shows

Figure 1.9 A footbridge with cable stayed mid-span support, Southampton, UK.

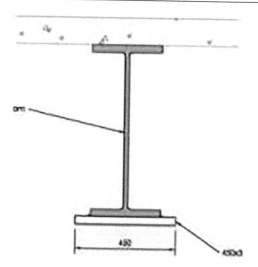

Figure 1.10 Increasing I value by adding a plate at the bottom of a beam.

such an example where cable stays to one side of the footbridge act as additional supports to reduce the deflection of the deck.

2. *Increasing second moment of area I:* This is normally applicable to individual members, such as by using a larger cross-section or adding material as far away as possible from the neutral axis of a given cross-section to enlarge the I value effectively. Figure 1.10 shows the familiar example of a long-narrow steel plate welded to the bottom of an I section steel beam. As the additional material was placed as far away from the neutral axis of the cross-section as possible, it effectively contributed to the second moment of area of the cross-section and resolved a possible vibration problem. Conceptually, a tall building can be seen as a large cantilever, the second moment of area of its cross-section can be increased by arranging the positions of columns, shear walls and bracing members of the building to be as far away as possible from the neutral axis of the cross-section.

3. *Reducing α:* This can be achieved by enhancing the boundary conditions, such as changing pinned supports to fixed supports. Alternatively, adding elastic supports to a structure can be adopted. For example, the cables of a cable-stayed bridge provide elastic supports to the bridge deck, allowing the bridge to span longer distances. In this case the bridge deck can be seen as a beam on an elastic foundation. The cable support shown in Figure 1.9 can also be explained as an elastic support.

Equation 1.15 is derived from simple beam theory and is applicable to any problems that can be converted to an equivalent beam, such as bridges or tall buildings. Actually, the understanding gained from this equation has been applied to more complex situations, such as plates, floors and roofs, far beyond beams. Many physical measures have been developed based on the three rules of thumb to design structures and structural members against deflection.

Equation 1.15 may relate to a single member or structural element as it is about the bending of a beam, but its application can have wider significance to engineering practice. It is considered that a similar equation at a whole-structure level would have even wider implementation in structural design against deflection.

1.5 Rules of Thumb for Design

There are "rules of thumb" for designing structural elements such as beams, columns and floors, which are simple and effective [1.1]. Such rules are familiar to most engineers and are widely used to develop quick preliminary designs. For example, for a given span and loading, they can quickly and sufficiently accurately indicate the required cross-section of a beam or the thickness of a reinforced concrete floor without calculation. These rules of thumb help not only speed up preliminary designs but also avoid mistakes.

The development of such rules of thumb can be illustrated by Figure 1.11. The rules of thumb are summarised from or can be abstracted from, 1) sound engineering practice and 2) subsequently based on, or checked by, theory. The rules can then be 3) used by many engineers in their design of structural elements such as beams, columns, walls and floors.

The usefulness of these rules of thumb and their development poses a question: are there other types of "rules of thumb" that can be used for designing whole structures to achieve smaller deflections, or for making structures more effective, efficient and even more elegant? For determining such rules, a logical way is to identify them through studying highly praised structures and reading the books written by the most eminent architects and engineers. After doing so it is possible to gain a higher level of appreciation of these structures, greater admiration from the heart of the creative designs and a deeper philosophical thinking of the relationships between form and function, between architecture and structures, and between art and technology, *etc.* However, it is very difficult to find such rules of thumb explicitly expressed by the great engineers, and some general rules cannot be abstracted from their designs or books, which could be passed on to others for use in the design of different structures.

An alternative way to find such rules of thumb for whole structures can be developed from theory as illustrated in Figure 1.12.

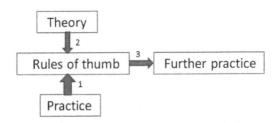

Figure 1.11 Development of rules of thumb to design structural elements.

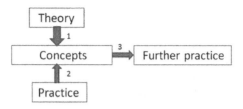

Figure 1.12 Development of structural concepts (rules of thumb) to design whole structures.

The flow chart in Figure 1.12 can be explained as follows:

1. Start to identify the concepts for a whole structure based on theory in current textbooks, which are significant for practical application in structural engineering. Here the words "structural concepts" are used instead of rules of thumb because they are general rather than specific, and can be used for designing many structures, and because particular physical measures need to be developed based on these structural concepts. Four structural concepts based on the relationships between deflection and internal forces of a structure are identified in Chapter 2. They can be intuitively interpreted and expressed in a concise and memorable manner as follows:

   - *The more direct the internal force paths, the smaller the deflection.*
   - *The smaller the internal forces, the smaller the deflection.*
   - *The more uniform the distribution of internal forces, the smaller the deflection.*
   - *The more the bending moments being converted into axial forces, the smaller the deflection.*

2. A number of hand calculation problems are examined with and without using one of the structural concepts. The outcomes demonstrate that the four structural concepts are both effective and efficient. Having used these structural concepts to examine several structures designed by well-known engineers and architects, it is fascinating to note that these structural concepts have been actually embedded in their designs! This explains why these structures are excellent from the structural point of view and it is observed that a structure is likely to be effective (smaller deflections), efficient (using less material) and elegant (architecturally pleasing) when one or more of the four structural concepts has been used.

3. It is hoped that these structural concepts, like some widely used rules of thumb for designing structural elements, can be used by many architects and engineers for designing structures against deflection and for achieving more effective, efficient and elegant designs.

1.6 Effectiveness, Efficiency and Elegance

In *The Structural Engineer*, the journal of the UK Institution of Structural Engineers, there was a definition of Structural Engineering on the contents page as follows [1.13].

Structural engineering is the science and art of designing and making, with economy and elegance, buildings, bridges, frameworks, and other similar structures so that they can safely resist the forces to which they may be subjected.

There are three key factors in the statement: *safety, economy* and *elegance* that can be seen as the objectives to be achieved in design and construction. The discipline of structural engineering allows structures to be produced with satisfactory performance at competitive costs. Elegance, which is not particularly related to safety and economy, is normally considered by architects.

For the purpose of this book, which focuses on the relationships between deflection and internal forces in structures, there is a need to scale down and revise the three objectives as effectiveness, efficiency and elegance. In general effectiveness means that a structure should satisfy all the functional requirements, such as those for deflection, stress and usage of the structure. Here effectiveness will be limited to deflection. If smaller maximum deflections are achieved in a design, it is likely, as discussed in Section 1.1, that the structure will have better buckling capacity, a higher fundamental natural frequency and smaller internal forces. Therefore, it can be said that this design is more effective than a similar design with a larger maximum deflection. Efficiency indicates the use of material in a design. When a structure is able meet the functional requirements using less material, it is said that this structure or design is more efficient than a similar one using more material. Elegance describes the pleasing and graceful visual appearance of the structure, which is perhaps somewhat subjective. Elegance here is considered to be structural elegance which results from structural correctness. For such definitions, the relative effectiveness and efficiency of two or more similar structures can be quantified.

The beauty and inspirational features of the four structural concepts to be studied in this book lies in that the effectiveness, efficiency and elegance of a structure are integrated as a whole. When one of the four concepts can be embedded into a design to make the structure more effective and efficient, it is likely that the structure will naturally become more elegant without purposely pursuing these aims. This point is demonstrated through a number of practical examples in Chapters 3 to 6.

1.7 Organisation of Contents

This book consists of seven chapters. The connections between the seven chapters are illustrated in Figure 1.13.

This chapter has provided an overview of the topic and the thoughts used to develop the contents of the book. Intuitive interpretation is emphasised in this book as it is an effective tool, and a skill, for reaching a higher level of understanding of structures, and this is further demonstrated in the later chapters.

Chapter 2 illustrates the theoretical background of the four structural concepts in an intuitive manner to enable the reader to gain a thorough understanding. The new meaning of the virtual work principle is explored and a basic equation, connecting deflection and internal forces of a whole structure, is

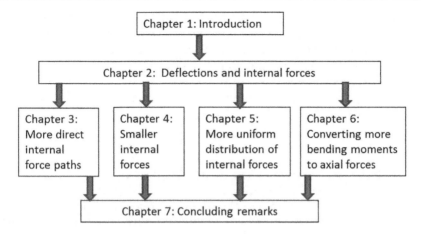

Figure 1.13 The connections between the chapters of this book.

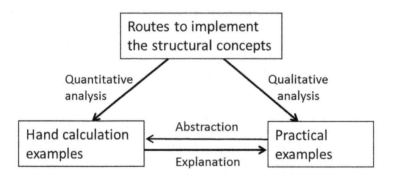

Figure 1.14 Connections between sections in Chapters 3–6.

intuitively interpreted. This leads to a new understanding of the four structural concepts on the relationships between smaller deflection and desirable distributions of internal forces, which provides a strong theoretical basis for wide and wise applications to be illustrated in the following chapters.

Chapters 3–6 are presented in a similar format and each of the chapters focuses on demonstrating the effectiveness and efficiency of one of the four structural concepts. Each of the chapters consists of three parts: 1) the routes to implementation of the structural concept in design, which are presented by physical and conceptual measures; 2) two hand calculation examples are provided which are abstracted from practical problems. Each example contains at least two similar cases, one with and one without involving an implementation measure based on the concept, by which the effect of the measure can be clearly identified and quantified; 3) several practical examples in which the implementation measures are effectively used are examined for demonstrating the application of the structural concept and its effects. The relationship between the three parts is shown in Figure 1.14. The detailed hand calculations will demonstrate the way of analysis for understanding and quantify the effectiveness and efficiency of the structural concept, and the findings will also serve for the comprehension of the related practical examples. The practical example will help the reader to realise how the concept has been used for the solution of challenging problems and for achieving more effective, efficient and elegant structures.

Chapter 7 provides concluding remarks and further discusses the use of the four concepts.

References

1.1   Schollar, T. *Structural Sizing: Rules of Thumb*, AJ, 1989.

1.2   Institution of Structural Engineers. *Dynamic Performance Requirements for Permanent Grandstands Subject to Crowd Action: Recommendations for Management, Design and Assessment*, The Institution of Structural Engineers, London, 2008.

1.3   Ellis, B. R. and Ji, T. *BRE Digest 426: The Response of Structures to Dynamic Crowd Loads*, Building Research Establishment Ltd., Watford, 2004.

1.4   Smith, A. L., Hicks, S. J. and Devine, P. J. *Design of Floors for Vibration: A New Approach*, The Steel Construction Institute, P354, Ascot, 2007.

1.5   Ji, T., Bell, A. J. and Ellis, B. R. *Understanding and Using Structural Concepts*, Second Edition, CRC Press, London, 2016.

1.6   Gere, J. M. *Mechanics of Materials*, Thomson Books/Cole, Belmont, 2004.

1.7   Hibbeler, R. C. *Mechanics of Materials*, Sixth Edition, Prentice-Hall Inc., Singapore, 2005.

1.8   Torroja, E. *The Structures of Eduardo Torroja: An Autobiography of an Engineering Accomplishment*, F W Dodge Corporation, USA, 1958.

1.9   Ji, T. and Bell, A. J. *Can Intuitive Interpretation Be Taught in Structural Engineering Education?* IV International Conference on Structural Engineering Education: Structural Engineering Education Without Borders, 20–22 June 2018, Madrid, Spain.

1.10 Ellis, B. R. and Ji, T. Human—Structure Interaction in Vertical Vibrations, *Structures and Buildings, the Proceedings of Civil Engineers*, 122(1), 1–9, 1997.

1.11 Meriam, J. L. and Kraige, L. G. *Engineering Mechanics, Vol. 2: Dynamics*, Fourth Edition, John Wiley & Sons, New York, 1998.

1.12 Ji, T. and Cunningham, L. S. An Insight into Structural Design Against Deflection, *Structures*, 15, 349–354, 2018.

1.13 The Institution of Structural Engineers. *The Structural Engineer*, 72(3), 1994.

Chapter 2

Deflections and Internal Forces

2.1 Deflection of a Structure

Equation 1.15 provides an explicit expression for the deflection of a beam at the structural element level, but its application extends far beyond structural elements. It is logical to examine similar equations which apply at the whole-structure level so that fundamental equations can be harnessed for more advanced design of structures against deflection.

At the structure level, the maximum deflections of any pin-connected structure and rigid frame structure with s members are shown in equations below [2.1, 2.2]:

$$\Delta_{max} = \sum_{i=1}^{s} \frac{N_i \bar{N}_i L_i}{E_i A_i} \tag{2.1}$$

$$\Delta_{max} = \sum_{i=1}^{s} \frac{\int_0^{L_i} M_i(x)\bar{M}_i(x)dx}{E_i I_i} \tag{2.2}$$

where N_i is the axial force in the ith member induced by the actual loads and $\bar{N}_i$ is the axial force in the ith member induced by a unit load applied at the critical point (location and direction) where the maximum deflection is likely to occur; $M_i(x)$ and $\bar{M}_i(x)$, similar to N_i and $\bar{N}_i$, are the bending moments in the ith member induced by the actual loads and by a unit load applied at the critical point respectively. L_i, E_i, A_i and I_i ($i = 1, 2, \ldots, s$) are the length, elastic modulus, area and second moment of area of the cross-section of the ith member.

Equations 2.1 and 2.2 provide a method for calculating the deflection of any framework structure with pinned or rigid connections. Equation 2.1 is suitable for trusses, scaffoldings and lattice structures, and has a history of over 150 years [2.3]. However, equation 2.1 has not been emphasised in textbooks on Mechanics of Materials and Structural Analysis to the same extent. This is because the use of the equation requires the calculation of the internal forces N_i and $\bar{N}_i$ to determine deflection, and such a calculation may be regarded as too tedious for structures with many members, or for statically indeterminate structures. Normally very simple statically determinate plane structures are provided in textbooks to show how equation 2.1 is used to calculate deflection. Similarly, equation 2.2 is used to calculate deflections of beams and simple frames.

Unlike equation 1.15, it is not obvious how to interpret equations 2.1 and 2.2 in a simple manner and to identify the physical essence embedded in the two equations. This is because N_i and $M_i(x)$ ($i = 1, 2, \ldots, s$) are functions of the loading that can have many variations and because equations 2.1 and 2.2 contain many items (*i.e.* a structure has many members).

In comparison with equation 1.15, the understanding and implementation of equations 2.1 and 2.2 for reducing the maximum deflection of a whole structure are not well known. Based on previous work [2.4–2.6], this chapter provides a theoretical basis to reveal the physical essence between the maximum deflection and the internal forces of a structure. The intuitive interpretation of the principle of virtual work will provide four fundamental structural concepts that are general, simple to understand, and are useful for practical applications.

2.2 Internal Forces, Deflections and Energies of Two Rods

The basic understanding of theory can often be established by studying simple cases. Figure 2.1a shows two linear elastic rods that have the same modulus E and the same length L but different cross-sectional areas, A_a and A_b where $A_b > A_a$ with $A_b = \alpha A_a$ ($\alpha > 1$) [2.2]. The thin and thick rods are subjected to two pairs of forces, P_a and P_b applied at their ends. Examine the relationships between the deflections, internal forces and elastic stain energies stored in the two rods for two loading conditions: 1) when the internal forces of the two rods are the same; and 2) when the total elongations of the two rods are the same.

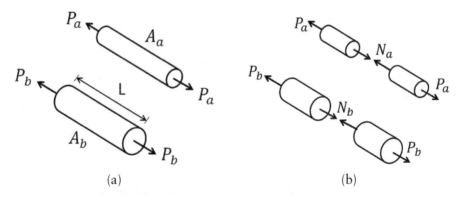

Figure 2.1 Two rods with different cross-sections undergoing axial deformation. (a) Two robs subjected to external axial loads. (b) Free body diagrams to reveal internal forces.

Using free-body diagrams (Figure 2.1b) and equilibrium equations, it is easy to determine that the internal forces in the two rods are equal to the external forces applied on them, $N_a = P_a$ and $N_b = P_b$.

The force-deflection relations for the two rods are:

$$\delta_a = \frac{N_a L}{EA_a} \text{ or } N_a = \frac{EA_a}{L}\delta_a = k_a \delta_a \tag{2.3a}$$

$$\delta_b = \frac{N_b L}{EA_b} \text{ or } N_b = \frac{EA_b}{L}\delta_b = k_b \delta_b \tag{2.3b}$$

where $k_a = EA_a / L$ and $k_b = EA_b / L$ are the axial stiffnesses of the two rods, and indicate the structural ability of the rods to resist axial deformation. The strain energies of the rods are respectively:

$$U_a = \frac{1}{2}k_a\delta_a^2 = \frac{1}{2}N_a\delta_a \text{ and } U_b = \frac{1}{2}k_b\delta_b^2 = \frac{1}{2}N_b\delta_b \tag{2.4a) and (2.4b}$$

Equation 2.3 indicates that *larger internal force will lead to larger deflection* while equation 2.4 shows that *larger deflection will lead to larger strain energy.* These statements come from the very simple case of a uniform rod, but they are applicable to more complex cases, even to whole structures.

The ratios of the two deflections in equation 2.3 and of the two energies in equation 2.4 are:

$$\frac{\delta_b}{\delta_a} = \frac{N_b L}{EA_b} \cdot \frac{EA_a}{N_a L} = \frac{A_a}{A_b}\frac{N_b}{N_a} = \frac{1}{\alpha}\frac{N_b}{N_a} \tag{2.5}$$

$$\frac{U_b}{U_a} = \alpha\frac{\delta_b^2}{\delta_a^2} = \frac{1}{\alpha}\frac{N_b^2}{N_a^2} \tag{2.6}$$

When the internal forces are the same for the two rods, *i.e.* $N_a = N_b$, it can be observed from equations 2.5 and 2.6 that $\delta_b < \delta_a$ and $U_b < U_a$ as $\alpha > 1$. *When the two rods are subjected to the same internal force, the thick rod has a smaller deflection and stores less energy than the thin rod.*

When the total deflections are the same for the two rods, *i.e.* $\delta_a = \delta_b$, it can also be seen from equations 2.5 and 2.6 that $N_b > N_a$ and $U_b > U_a$. This indicates that *when the two rods are made to deflect the same amount, the thick rod will experience larger internal forces and will store more strain energy than the thin rod.*

To demonstrate an implication of the last statement, the two rods are now used to support a weightless rigid plate which in turn supports a concentrated vertical load of P. To create a symmetric problem, the central rod has a cross-section area A_b and two side rods have cross-section areas of $A_a / 2$ (replacing the original single rod A_a) as shown in Figure 2.2a. Using the free-body diagram shown in Figure 2.2b, the equilibrium equation and the force-deflection equation give:

$$P = N_a + N_b = (\frac{EA_a}{L} + \frac{EA_b}{L})\Delta = (k_a + k_b)\Delta \qquad (2.7)$$

where Δ is the vertical deflection of the rods.

Equation 2.7 indicates that *the stiffer member shares or attracts a larger portion of the load for the same deflection,* or *the internal forces in the members are proportional to their axial stiffnesses.* This statement is derived from a simple axial compression problem, but it is applicable to more complex situations. For example, a weightless rigid plate is supported by four columns and is subjected to a concentrated lateral load as shown in Figure 2.3a. The four columns have the same height of L and the same elastic modulus E but different second moments of areas, I_a, I_b, I_c and I_d. This is a bending problem, and the four columns experience the same amount of lateral deflection. The

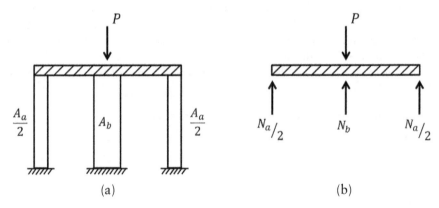

Figure 2.2 A compression problem. (a) A structure. (b) Free body diagram.

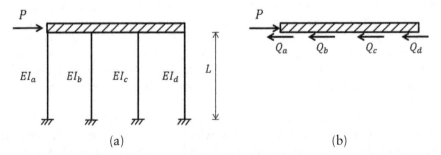

Figure 2.3 A bending problem. (a) A structure. (b) Free body diagram.

free-body diagram of the rigid plate is shown in Figure 2.3b and the equilibrium equation for the rigid plate is:

$$P = Q_a + Q_b + Q_c + Q_d = (\frac{12EI_a}{L^3} + \frac{12EI_b}{L^3} + \frac{12EI_c}{L^3} + \frac{12EI_d}{L^3})\Delta$$
$$= (k_a + k_b + k_c + k_d)\Delta$$

(2.8)

where $k_i = 12EI_i / L^3$ $(i = a,b,c,d)$. Equation 2.8 has a similar pattern to equation 2.7, and therefore the observation, or conclusion, from equation 2.7 is applicable to bending problems as described by equation 2.8. The force transmission from loading positions to structural supports can be seen as a force flow through the structural members to the supports, with the stiffer members attracting a larger force flow. A particular case is considered that $I_a = I_b = I_c$ and $I_d = 2I_a$, According to equation 2.8, the three left-hand columns attract $0.2P$ each and the right-hand column attracts $0.4P$. The result indicates that *force flows more to the stiffer parts of the structure* and the *force flow can thus be guided through design*.

2.3 Internal Forces, Deflection and Energy of a Structure

It is of practical importance to know the position of the critical point at which the maximum deflection of a structure is likely to occur. To identify such a point, a unit load can be placed in the appropriate direction at every nodal point in turn and its corresponding deflection calculated, leading to an array of nodal deflections. The nodal point at which the maximum value in the array corresponds to is the *critical point*. According to this definition, the critical point of a structure is independent of the loading on the structure. Normally, the critical point of a structure can be intuitively identified without calculations. For example, the critical point of a cantilever is at its free end and the critical point of a simply supported plate is at its centre. For the particular case in Figure 2.4, node C is the critical point of the truss in the vertical direction.

Consider a truss structure that consists of s members and n degrees of freedom which is subjected to two sets of loading as shown in Figure 2.4. All members of the truss have the same elastic modulus E. Load case 1, shown in Figure 2.4a, is the actual loading, $\{P_1\}$, and Load case 2, shown in Figure 2.4(b), is a specific loading case $\{P_2\}$ in which a unit concentrated load is applied at the critical point C of the structure.

Analysing the two truss structures leads to two sets of results in which subscripts 1 and 2 respectively relate to the Load case numbers.

Load case 1: There are external and internal forces $\{P_1\}$ and $\{N_1\}$, the nodal deflections are $\{\Delta_1\}$ and the member elongations are $\{\delta_1\}$. The relationship between the internal force and the elongation of the jth member is $\delta_{1,j} = N_{1,j}L_j / EA_j$, where L_j and A_j are the length and the cross-sectional area of the jth member.

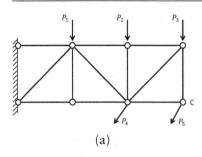
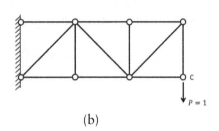

(a)   (b)

Figure 2.4 Two sets of loading on the same truss structure (a) Load case 1; (b) Load case 2.

Load case 2: There are similar quantities, $\{P_2\}$, $\{N_2\}$, $\{\Delta_2\}$ and $\{\delta_2\}$, and the relationship $\delta_{2,j} = N_{2,j}L_j / EA_j$.

For a conservative system, the work-energy principle states that *if the stresses in a body do not exceed the elastic limit, all the work done on a body by external forces is equal to the elastic strain energy stored in the body* [2.2], which can be expressed for Case 1 as:

$$W_{1,1} = \frac{1}{2}\sum_{i=1}^{n} P_{1,i}\Delta_{1,i} = \frac{1}{2}\sum_{j=1}^{s} N_{1,j}\delta_{1,j} = \frac{1}{2}\sum_{j=1}^{s} \frac{N_{1,j}^2 L_j}{EA_j} \tag{2.9}$$

where $W_{1,1}$ is the external work done by the loads $\{P_1\}$ on the deflections $\{\Delta_1\}$ induced by $\{P_1\}$ and the right-hand side of equation 2.9 is the elastic energy stored in the s members.

Consider the work done by the loads $\{P_1\}$ in Case 1 moving through the deflections $\{\Delta_2\}$ resulting from the loads in Case 2, and the strain energy created by the internal forces $\{N_1\}$ in Case 1 acting on the member elongations $\{\delta_2\}$ in Case 2. This leads to:

$$W_{1,2} = \frac{1}{2}\sum_{i=1}^{n} P_{1,i}\Delta_{2,i} = \frac{1}{2}\sum_{j=1}^{s} N_{1,j}\delta_{2,j} = \frac{1}{2}\sum_{j=1}^{s} \frac{N_{1,j}N_{2,j}L_j}{EA_j} \tag{2.10}$$

Similarly, the work done by the loads $\{P_2\}$ moving through the deflections $\{\Delta_1\}$ and the strain energy contributed by the internal forces $\{N_2\}$ on the member elongations $\{\delta_1\}$ are:

$$W_{2,1} = \frac{1}{2}\sum_{i=1}^{n} P_{2,i}\Delta_{1,i} = \frac{1}{2}\sum_{j=1}^{s} N_{2,j}\delta_{1,j} = \frac{1}{2}\sum_{j=1}^{s} \frac{N_{2,j}N_{1,j}L_j}{EA_j} \tag{2.11}$$

It can be observed that the right-hand side items in equations 2.10 and 2.11 are the same, which leads to:

$$W_{1,2} = W_{2,1} \tag{2.12}$$

This is the *reciprocal theorem of work*. It states that *the work done by the loads in Case 1 moving through the deflections resulting from the loads in Case 2 is equal to the work done by the loads in Case 2 moving through the deflections induced by the loads in Case 1.*

As only a unit load is applied at node C in Case 2, the external work in equation 2.11 becomes:

$$W_{2,1} = \frac{1}{2}\sum_{i=1}^{n} P_{2,i}\Delta_{1,i} = \frac{1}{2}(1 \times \Delta_{1,C}) \tag{2.13}$$

Substituting equation 2.13 into equation 2.11 and simplifying gives:

$$\Delta_{1,C} = \sum_{j=1}^{s} \frac{N_{2,j}N_{1,j}L_{j}}{EA_{j}} \tag{2.14}$$

Equation 2.14 provides a method for calculating the deflection at node C of the structure resulting from the loads in Case 1 (Figure 2.4a) in three steps:

1. Calculate the internal forces $\{N_1\}$ resulting from $\{P_1\}$, which are the actual loads on the structure.
2. Calculate the internal forces $\{N_2\}$ resulting from the unit load $\{P_2\}$.
3. Use equation 2.14 to calculate the vertical deflection at node C.

It is noted that the calculations require to determine $\{N_1\}$ and $\{N_2\}$, which can be challenging and tedious if a truss structure is statically indeterminate or if it has many members. Hence, equation 2.14 is seldom used to calculate the deflections of actual truss structures.

It is logical to examine $W_{2,2}$ after examining $W_{1,1}$, $W_{1,2}$ and $W_{2,1}$. $W_{2,2}$ can be expressed as:

$$W_{2,2} = \frac{1}{2}(1 \times \Delta_{2,C}) = \frac{1}{2}\sum_{j=1}^{s} \frac{N_{2,j}^{2}L_{j}}{EA_{j}} \tag{2.15}$$

This is a similar equation for $W_{1,1}$, but $W_{2,2}$ means the work done by $\{P_2\}$ moving through the deflection $\{\Delta_2\}$, *i.e.* by a unit force $P_{2,C} = 1$ moving through the deflection $\Delta_{2,C}$. Simplifying Equation 2.15 gives:

$$\Delta_{2,C} = \sum_{j=1}^{s} \frac{N_{2,j}^{2}L_{j}}{EA_{j}} \tag{2.16}$$

Before discussing the physical meaning of equation 2.16, a beam type of structure is considered in which bending moments are the major internal forces and to which similar equations to those for truss structures apply. If the pinned connections of the truss in Figure 2.4 are all changed to rigid connections, it becomes a frame structure. The two equations for calculating the deflections due to the actual loads and the deflections due to the unit load, considering bending moments alone, can be written as:

$$W_{21} = \frac{1}{2}\sum_{i=1}^{n}P_{2,i}\Delta_{1,i} = \frac{1}{2}(1 \times \Delta_{1,C}) = \frac{1}{2}\sum_{j=1}^{s}\frac{\int_{0}^{L_j} M_{2,j}(x)M_{1,j}(x)dx}{EI_j} \qquad (2.17)$$

$$W_{22} = \frac{1}{2}\sum_{i=1}^{n}P_{2,i}\Delta_{2,i} = \frac{1}{2}(1 \times \Delta_{2,C}) = \frac{1}{2}\sum_{j=1}^{s}\frac{\int_{0}^{L_j} M_{2,j}^2(x)dx}{EI_j} \qquad (2.18)$$

where $M_{1,j}(x)$ and $M_{2,j}(x)$ are the bending moments along the jth member induced by Load cases 1 and 2 respectively. Similar to equation 2.14, equation 2.17 can be used to calculate the deflection of a frame structure. The integration $\int_{0}^{L_j} M_{2,j}^2(x)dx$ for the jth member in equation 2.18 means the area under the curve for $M_{2,j}^2(x)$ between 0 and L_j, which can also be represented by the same area of an equivalent rectangle with a length L_j and a mean height $\bar{M}_{2,i}^2$. Equation 2.18 can therefore also be stated as:

$$\Delta_{2,C} = \sum_{j=1}^{s}\frac{\bar{M}_{2,j}^2 L_j}{EI_j} \qquad (2.19)$$

Equations 2.16 and 2.19 have a similar form involving an internal force squared with one equation applicable to truss structures and one applicable to beam and frame structures. The physical meaning of $\Delta_{2,C}$ in the two equations will be examined in the next section.

2.4 Physical Meaning of $\Delta_{2,C}$

Consider the most unfavorable loading scenario that all the loads on a structure are lumped at the critical point. This leads to the largest deflection at the point of those induced by all possible loading distributions. For example, all the loads acting on the truss in Figure 2.4a are moved to and lumped at point C, the vertical deflection at C due to the lumped loads will be larger than those induced by any other loading distributions. If this lumped load is then normalised to a unit load, which is not a true loading condition, but is the worst load case for the maximum deflection of a structure, equations 2.16 or 2.19 can be used to calculate the normalised maximum deflections of different types of truss and frame structures. Therefore $\Delta_{2,C}$ *means the maximum deflection under the most unfavourable loading scenario in which all the loads on a structure are lumped at the critical point and normalised to a unit.*

The flexibility coefficient at a point of a structure is defined as the deflection induced by a unit load in the loading direction. Therefore $\Delta_{2,C}$ (equation 2.19) is the flexibility coefficient at the critical point of the structure and has the largest value among all flexibility coefficients for any truss or frame structure. This interpretation can be demonstrated mathematically.

Considering a structure that is modelled by s elements and n nodes with each node having d degrees of freedom. The static equilibrium equation, containing $n \times d$ unknowns, is expressed as:

$$[K]\{U\} = \{P\} \tag{2.20}$$

where $\{U\}$ is the nodal displacement vector to be determined, $\{P\}$ is the load vector and $[K]$ is the stiffness matrix that includes the effect of the boundary conditions. Equation 2.20 is a general equation of equilibrium and is suitable for any linear elastic structure.

When a single unit load is applied at the critical node C in a given direction, l, the lth degree of freedom of a node, the load vector is:

$$\{P\} = \{0,\ 0, \ldots 1, \ldots, 0, 0\}^{\mathrm{T}} \tag{2.21}$$

Substituting equation 2.21 into equation 2.20 and solving leads to the deflections [2.5]:

$$
\begin{Bmatrix} u_1 \\ \vdots \\ u_{cl} \\ \vdots \\ u_{n \times d} \end{Bmatrix} = [K]^{-1}\{P\} = [\delta]\{P\} =
\begin{bmatrix} \delta_{1,1} & \cdots & \delta_{1,cl} & \cdots & \delta_{1,n} \\ \vdots & \ddots & \vdots & \ddots & \vdots \\ \delta_{cl,1} & \cdots & \delta_{cl,cl} & \cdots & \delta_{cl,n} \\ \vdots & \ddots & \vdots & \ddots & \vdots \\ \delta_{n,1} & \cdots & \delta_{n,cl} & \cdots & \delta_{n,n} \end{bmatrix}
\begin{Bmatrix} 0 \\ \vdots \\ 1 \\ \vdots \\ 0 \end{Bmatrix} =
\begin{Bmatrix} \delta_{1,cl} \\ \vdots \\ \delta_{cl,cl} \\ \vdots \\ \delta_{n,cl} \end{Bmatrix} \tag{2.22}
$$

where $[\delta]$ is the flexibility matrix of the structure (the inverse of the stiffness matrix), and $\delta_{cl,cl}$ is the diagonal element at row cl and column cl in the flexibility matrix, in which cl means the critical degree of freedom and can be determined by $cl = (c-1) \times d + l$ where c is the node number of the critical point C, d is the degrees of freedom of each node and l is the number of the concerned degree of freedom. Considering the row cl in equation 2.22 gives:

$$\Delta_{2,C} = u_{cl} = \delta_{cl,cl} \tag{2.23}$$

Equation 2.23 states that *the deflection at the critical point u_{cl} ($\Delta_{2,C}$) induced by a unit load at this degree of freedom is the coefficient $\delta_{cl,cl}$ in the flexibility matrix of the structure.*

The stiffness matrix $[K]$ gives a detailed description of the distribution of structural members and their contribution to the stiffness matrix. However, from $[K]$ it is difficult to sense how stiff the structure is. In practice, a unique value is preferred to define the stiffness of a structure. It is common for the inverse of the deflection at the critical point induced by a unit load is defined as the *static stiffness* of the structure [2.6], *i.e.*:

$$K_S = \frac{1}{u_{cl}} \tag{2.24}$$

For example, when a unit vertical downward load is applied at the free end of a cantilever, the vertical deflection at the free end is $L^3 / (3EI)$, the static stiffness of the cantilever is then $(3EI) / L^3$. Equations 2.23 and 2.24 give the static stiffness of the structure as:

$$K_S = \frac{1}{\delta_{cl,cl}} \tag{2.25}$$

Equation 2.25 indicates that *the static stiffness of a structure is the inverse of the largest diagonal element in the flexibility matrix of the structure.*

In summary, *the physical meaning of* $\Delta_{2,C}$ *(the deflection at the critical point of a structure due to a unit load applied at this point) is the inverse of the static stiffness of a structure (equation 2.24) and is equal to the largest flexibility coefficient in the flexibility matrix of the structure (equation 2.23).* $\Delta_{2,C}$ can also be seen as the largest possible deflection when all the loads are lumped to the critical point of the structure and scaled to a unit load.

2.5 Intuitive Interpretation

After examining the physical meaning of the left sides of equations 2.16 and 2.19, it is possible to interpret the right sides of the two equations. As the loading has been lumped at the critical point of the structure and normalised to a unit load, the internal forces in equations 2.16 and 2.19 are independent of any particular loading but are functions of structural form. For statically indeterminate structures they are also functions of material and cross-sectional properties. For the purpose of design, it is ideal to make the deflections of a structure as small as possible, or the static stiffness of the structure as large as possible, using the same amount of material or less material, *i.e.*:

$$\frac{1}{K_S} = \Delta_{2,C} = \sum_{j=1}^{s} \frac{N_{2,j}^2 L_j}{EA_j} \to \min \tag{2.16}$$

$$\frac{1}{K_S} = \Delta_{2,C} = \sum_{j=1}^{s} \frac{\bar{M}_{2,j}^2 L_j}{EI_j} \to \min \tag{2.19}$$

Finding the minimum deflection at the critical point or the largest static stiffness of a structure may be considered as a topology optimisation problem. For one type of topology optimisation [2.7], the geometry of a structure is altered gradually by removing the element with the smallest stress or adding an element where the stress demand is high. This iterative process seeks to make the distribution of stress as uniform as possible and eventually leads to an optimum topology design based on the objective function, a stiffer structure. Equation 2.16 or equation 2.19 forms an incompletely defined optimisation problem, and therefore standard optimisation techniques may not be directly applicable at this stage. However, the physical essence of the incomplete optimisation problem can still be identified and interpreted.

As internal forces and structural form are closely related, the internal forces can be examined directly using equations 2.16 and 2.19, instead of considering the topology of the structure. The physical quantities in equations 2.16 and 2.19 have the following mathematical characteristics:

1. $E > 0; A_j > 0; I_j > 0$ and $L_j > 0$;

2. $N_{2,j}^2 \geq 0$ and $\overline{M}_{2,j}^2 \geq 0$

regardless of whether the member is in tension or in compression or whether the bending moment is positive or negative.

All the items in equations 2.16 and 2.19 are positive or zero, *i.e.* there are no negative terms. When A_j/L_j in equation 2.16 and I_j/L_j in equation 2.19 do not change significantly, the internal forces dominate the deflections in the two equations. The relationships between the smaller deflections and internal forces embedded in the two equations can be interpreted intuitively as follows:

1. One way to make the deflection as small as possible is to have as many terms as possible equal to zero on the right sides of equations 2.16 and 2.19. Mathematically, the fewer the positive terms, the smaller the sum of all the terms. Physically, many zero terms means that these members are zero-force members. The unit load positioned at the critical point is transmitted to the supports of the structure without passing through these zero-force members and takes a shorter internal force path. The greater the number of zero-force members, the more direct the internal force path. This physical phenomenon suggests that *shorter or more direct internal force paths from the load to the structural supports lead to smaller deflection of a structure.*

2. It can be directly observed from equations 2.16 and 2.19 that the sums will be smaller if each of the terms becomes smaller. The corresponding physical phenomenon is that *smaller internal forces lead to smaller deflections of a structure.*

3. Consider three sets of data, each consisting of five numbers as shown in Table 2.1. The sums of the three sets of data are the same, but the largest differences between the five numbers in the three data sets are different. Consequently, the sums of the squares of the data in the three sets are different. It can be observed that the larger the difference of the five numbers, the larger the sum of the squares.

Table 2.1 Comparison of Three Sets of Data

Data set	Five data	Largest difference in the five data	$\sum_{i=1}^{5} a_i$	$\sum_{i=1}^{5} a_i^5$
.1	1, 2, 3, 4, 5	4	15	55
2	2, 2, 3. 4, 4,	2	15	49
3	3, 3, 3, 3, 3	0	15	45

Due to the similarity between the right-hand side of equation 2.16 or 2.19 and $\sum_{i=1}^{5} a_i^2$, the observation from the simple comparison in Table 2.1 is applicable to equations 2.16 and 2.19. Smaller differences between the internal forces will lead to a smaller sum of squares than those with larger differences. This can be interpreted physically as: ***more uniformly distributed internal forces result in smaller deflections in a structure.***

In summary, there are three ways to achieve smaller deflections by actively achieving desirable internal forces and force distributions. They can be presented in a more memorable way as follows:

1. The more direct the internal force paths, the smaller the deflection of a structure;
2. The smaller the internal forces, the smaller the deflection of a structure;
3. The more uniform the distribution of internal forces, the smaller the deflection of a structure.

2.6 Deflections due to Bending Moment, Axial and Shear Forces

The previous interpretation of how to achieve smaller deflections is based on equations 2.16 and 2.19, which are based on either axial forces or bending moments. It is however possible that members can be subjected to bending moment, axial force and shear force simultaneously. For structures containing such members the deflection of a structure is expressed as:

$$\Delta_C = \sum_{i=1}^{s} \frac{\int_0^{L_i} M_i(x)\bar{M}_i(x)dx}{E_i I_i} + \sum_{i=1}^{s} \frac{N_i(x)\bar{N}_i(x)L_i}{E_i A_i} + \sum_{i=1}^{s} \frac{Q_i(x)\bar{Q}_i(x)l}{G_i A_i} \qquad (2.26)$$

The deflections contributed by bending, axial and shear effects can be illustrated by an example. Consider a quarter of a circular ring with a radius of R, one fixed end, and a free end, as shown in Figure 2.5. The curved member has a uniform rectangular cross-section with width b and height h and material properties of E and $G = 0.5E$. A unit downward load is applied at the free end of the member. Determine the vertical deflections at the free end of the member.

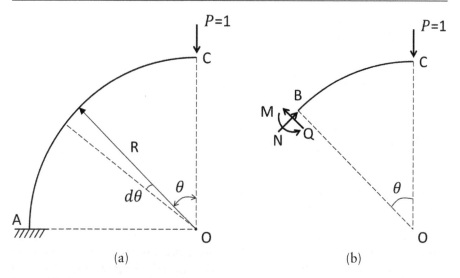

Figure 2.5 A quarter ring subjected to a concentrated load. (a) The ring structure. (b) Free body diagram.

Consider a free body diagram of part of the ring as shown in Figure 2.5b, which shows the internal forces at a typical cross-section B, defined by θ. The internal forces in the member can be determined using three basic equilibrium equations:

$$M = \bar{M} = R\sin\theta; \; N = \bar{N} = \sin\theta; \; Q = \bar{Q} = \cos\theta$$

Substituting the internal forces into equation 2.26 and noting that $dl = Rd\theta$, gives:

$$\Delta = \frac{\int_0^{L_i} M(x)\bar{M}(x)dl}{EI} + \frac{\int_0^{L_i} N(x)\bar{N}(x)dl}{EA} + \frac{\int_0^{L_i} Q(x)\bar{Q}(x)dl}{GA}$$

$$= \frac{R^3}{EI}\int_0^{\pi/2}\sin^2\theta d\theta + \frac{R}{EA}\int_0^{\pi/2}\sin^2\theta d\theta + \frac{R}{GA}\int_0^{\pi/2}\cos^2\theta d\theta$$

$$= \frac{\pi R^3}{4EI} + \frac{\pi R}{4EA} + \frac{\pi R}{4GA}$$

Now substituting $G = 0.5E$ and $A = 12I/h^2$ into the previous equation, the vertical deflection at the free end is:

$$\Delta = \frac{\pi R^3}{4EI}[1 + \frac{1}{12}\left(\frac{h}{R}\right)^2 + \frac{1}{6}\left(\frac{h}{R}\right)^2] \tag{2.27}$$

The terms in the square bracket show the relative contributions to the deflection from bending moment, axial force and shear force. The relative contributions can be quantified for three values of the ratio of the cross-section height to the radius of the curved member as follow:

For $\dfrac{h}{R} = 10$ $\Delta = \dfrac{\pi R^3}{4EI}[1 + \dfrac{1}{1200} + \dfrac{1}{600}]$

For $\dfrac{h}{R} = 5$ $\Delta = \dfrac{\pi R^3}{4EI}[1 + \dfrac{1}{300} + \dfrac{1}{150}]$

For $\dfrac{h}{R} = 2.5$ $\Delta = \dfrac{\pi R^3}{4EI}[1 + \dfrac{1}{75} + \dfrac{1}{37.5}]$

The contributions from the axial and shear actions are very small in comparison to that arising from bending. **When the dimensional sizes of a member are significantly larger than its cross-sectional sizes, the deflections induced by axial and shear actions in a bending problem are very small and can be neglected.**

For a structure subjected to bending and axial actions with f members subjected to bending and g members subjected to axial force, the deflection of the structure can be determined from equations 2.16 and 2.19 as:

$$\Delta_{2,C} = \sum_{j=1}^{f} \frac{\bar{M}_{2,j}^2 L_j}{E_j I_j} + \sum_{j=1}^{g} \frac{N_{2,j}^2 L_j}{E_j A_j} \tag{2.28}$$

As the deflection induced by bending action is much larger than that induced by axial force action, equation 2.28 implies another way to reduce the deflection **by converting bending moment actions into axial force actions,** for example by replacing bending members with axial force members and/or by adding bar members to reduce bending members in a structure.

This can be presented as the fourth structural concept to achieve smaller deflection as follows:

4. **The more the bending moments are converted into axial forces, the smaller the deflection of a structure.**

It is well understood that structures will become more efficient when loads are transmitted through axial forces rather than bending moments. One of the reasons is to achieve the efficiency of materials, which relates to the stress distributions on the cross-sections of members, *i.e.* a uniform distribution for axial forces and a linear distribution for bending moment. The fourth structural concept is particularly related to deflection of a structure and indicates the deflection induced by bending moments will be much larger than that by axial forces.

2.7 Characteristics of the Structural Concepts

2.7.1 The Four Structural Concepts

The four structural concepts intuitively interpreted from equations 2.16, 2.19 and 2.28 are simple, meaningful, fundamental and general, and they are related to the deflections and internal forces of a whole structure that can be any type of truss and/or frame structure. These four structural concepts can be summarised in a more concise and memorable manner and treated as rules of thumb as follows:

1. *The more direct the internal force paths, the smaller the deflection.*
2. *The smaller the internal forces, the smaller the deflection.*
3. *The more uniform the distribution of internal forces, the smaller the deflection.*
4. *The more the bending moments are converted into axial forces, the smaller the deflection.*

In these statements, the form of a structure is not explicitly stated but is embedded. It has been shown in Section 1.2 that structural form, deflection and internal forces are closely related so that altering any one of the three will lead to a change of the other two. The four structural concepts provide a solid basis for creative applications. They will be examined and discussed further to gain a sound and thorough understanding.

2.7.2 Generality

Equations 2.16 and 2.19 are derived from the principle of virtual work and are general and applicable to all types of truss and frame structures and include the structural concepts derived from equation 1.15 which are based on beam theory.

The maximum bending moment of a uniform beam subjected to a uniformly distributed load is:

$$M_{max} = \beta q L^2$$

(2.29)

For a simply supported beam, $\beta = 1/8$, and for a cantilever, $\beta = 1/2$. Substituting equation 2.29 into equation 1.15, the deflection can be alternatively expressed as:

$$\Delta_{max} = \alpha \frac{qL^4}{EI} = \alpha \frac{M_{max}^2}{\beta^2 qEI}$$

(2.30)

Equation 2.30 states that *the maximum deflection is proportional to the maximum bending moment squared* or in more general terms, *the smaller the internal forces, the smaller the deflection,* which is the second structural concept. This demonstrates that the four structural concepts derived using the principle of virtual work for a whole structure include the basics developed from beam theory.

2.7.3 Interchangeability

The first three structural concepts are abstracted from the same equations (equations 2.16 and 2.19), which means that these structural concepts are not independent and are exchangeable, *i.e.* if a structure reaches a state with a more direct internal force path, it is likely that the structure will have smaller internal forces and a more uniform distribution of internal forces. This can be illustrated using an example.

Figure 2.6 shows two similar 3-bay and 3-storey truss type structures carrying a unit horizontal load at the top right corner. They have the same dimensions, the same material property, E, and cross-sectional area, A. There are 24 members in each frame and the horizontal and vertical members have the same length of L. The only difference between the two frames is the arrangement of the three bracing members. For Frame A in Figure 2.6a, the bracing members are placed in the right bay and for Frame B, the bracing members are arranged diagonally across all three bays of the structure. The bracing arrangement in Frame B can be evolved from that in Frame A by moving the middle bracing member one panel to its left and the bottom bracing member two panels to the left. The two frames are statically determinate, and their internal forces can be easily calculated by hand. For the convenience of the comparison, the non-zero internal forces are indicated next to the corresponding members of the two frames in Figure 2.6.

The horizontal deflections at the loading positions of Frames A and B can be calculated using equation 2.16 as follows:

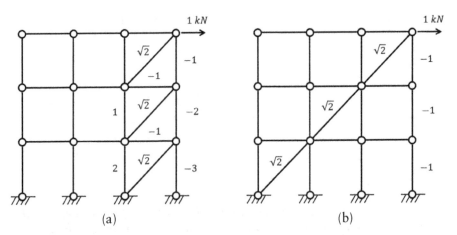

(a)   (b)

Figure 2.6 Two 3-bay and 3-storey frames. (a) Frame A. (b) Frame B.

$$\Delta_A = \sum_{j=1}^{s} \frac{N_j^2 L_j}{EA} = \frac{L}{EA}[4 \times (1)^2 + 2 \times (2)^2 + (3)^2 + 3(\sqrt{2})^2 \sqrt{2}]$$

$$= \frac{L}{EA}[4 + 8 + 9 + 6\sqrt{2}] = \frac{(21 + 6\sqrt{2})L}{EA} = \frac{29.49L}{EA}$$

(2.31)

$$\Delta_B = \sum_{j=1}^{s} \frac{N_j^2 L_j}{EA} = \frac{L}{EA}[3 \times (1)^2 + 3(\sqrt{2})^2 \sqrt{2}] = \frac{(3 + 6\sqrt{2})L}{EA} = \frac{11.49L}{EA}$$ (2.32)

The ratio of the two deflections is:

$$\frac{\Delta_B}{\Delta_A} = \frac{11.49}{29.49} = 0.39$$ (2.33)

The deflection of Frame B is only 39% of that of Frame A with the same amounts of material used.

The reasons that the deflection of Frame B is much smaller than that of Frame A can be explained intuitively using the first three structural concepts. It is observed from Figure 2.6 that:

- Ten members have internal forces in Frame A while six members have internal forces in Frame B indicating that Frame B creates more direct internal force paths to transmit the load to its supports than Frame A (Structural Concept 1), which leads to a smaller deflection with over 60% reduction.
- The largest force has a magnitude of 3 in Frame A while it is $\sqrt{2}$ in Frame B, *i.e.* Frame B has smaller internal forces than Frame A (Structural concept 2).
- The maximum difference between internal forces is $|3| - |1| = 2$ in Frame A while the difference is $|\sqrt{2}| - |1| = 0.414$ in Frame B. This indicates that Frame B has a more uniform distribution of the internal forces than Frame A (Structural concept 3).

It can be observed from this example that the first three structural concepts are exchangeable. Although any of the three structural concepts can be used for the design of the bracing patterns, for this particular example achieving a more direct internal force path is easier than creating smaller internal forces or a more uniform distribution of internal forces. In other cases, using the second or third structural concepts may be more convenient than using the first structural concept. This understanding is useful for design as different approaches can be followed to achieve smaller deflections.

2.7.4 Compatibility

The first three structural concepts may not be fully compatible as they are stated from different perspectives based on the same equations. A more direct internal force path requires that more members are in a zero-force state which may lead to larger internal forces in the other members. On the other hand, the more uniform distribution of internal forces may imply that more members share internal forces so that there are no large differences between the internal forces in individual members. This type of incompatibility can also be demonstrated using an example.

Two similar 3-bay and 4-storey truss type structures with the same dimensions are shown in Figure 2.7. The horizontal and vertical members have the same length of L and all members have the same material property E and cross-sectional area A. Each frame has 32 members including 4 bracing members. The only difference between the two frames is the arrangement of the bracing members in the bottom left panels.

> **Frame A:** The bottom bracing member is placed between nodes B and D and is linked with the bracing member in the upper storey.
> **Frame B:** The bottom bracing member is linked between nodes A and C and is parallel to that in the upper storey.

The two frames are statically determinate, and their internal forces can be easily calculated by hand and the non-zero internal forces are indicated next to the corresponding members in Figure 2.7. It can be observed from Figure 2.7 that only nine members are in a non-zero force state in Frame A while eleven members are in a non-zero force state in Frame B, indicating that Frame A creates a more direct internal force path than that in Frame B. However, there are smaller internal forces and smaller differences between the internal forces in Frame B than that in Frame A, indicating that Frame B creates a more uniform distribution of internal forces. Which frame has a smaller deflection? Equation 2.16

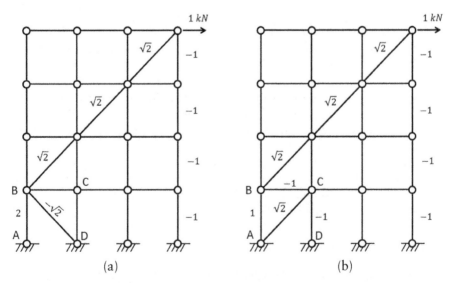

(a)　　　　　　　　　　　　　(b)

Figure 2.7 Comparison of internal forces of the two simple frames. (a) Frame A. (b) Frame B.

can be used to determine the deflections of the two frames with the internal forces indicated in Figure 2.7:

$$\Delta_A = \sum_{j=1}^{s} \frac{N_j^2 L_j}{EA} = \frac{L}{EA}[4 \times (1)^2 + (2)^2 + 4(\sqrt{2})^2 \sqrt{2}]$$

$$= \frac{(8 + 8\sqrt{2})L}{EA} = \frac{19.31L}{EA}$$

(2.34)

$$\Delta_B = \sum_{j=1}^{s} \frac{N_j^2 L_j}{EA} = \frac{L}{EA}[7 \times (1)^2 + 4(\sqrt{2})^2 \sqrt{2}] = \frac{(7 + 8\sqrt{2})L}{EA} = \frac{18.31L}{EA} \qquad (2.35)$$

It can be seen that Frame B has a smaller deflection than Frame A, although it has less direct internal force paths than Frame A. Comparing the internal forces in Frames A and B (Figure 2.7), only three members, AB, BC and CD, have different internal forces, which makes the difference between the calculated horizontal deflections. For Frame A, member AB has an internal force of 2kN and the two other members have zero-force, while for Frame B, the three members have the same force magnitude of 1kN. Due to the action of "square," $2^2 > 3 \times 1^2$, *i.e.* the contribution of the internal force in member AB in Frame A to the deflection is larger than that from the three members in Frame B. In this example, the structural concepts of smaller internal forces and a more uniform distribution of internal forces are even more effective than the structural concept of a direct internal force path.

This example shows that the first three structural concepts are not fully compatible and also tells that there are opportunities for creative use of the structural concepts.

2.7.5 Reversibility

The presentation of the four structural concepts seems to indicate that a smaller deflection is the consequence of more direct internal force paths, smaller internal forces, more uniform distribution of internal forces or converting bending moments to axial forces. As deflection and internal forces occur at the same time when a structure is loaded, the structural concepts can also be stated in reverse as:

1. The smaller the deflection, the more direct the internal force path.
2. The smaller the deflection, the smaller the internal forces.
3. The smaller the deflection, the more uniform the distribution of internal forces.
4. The smaller the deflection, the more the bending moments are converted into axial forces.

These reverse statements say that internal forces can be reduced or more uniformly distributed when the deflection of a structure can be controlled or

Figure 2.8 Additional supports to a floor.

reduced. Controlling or reducing deflection where possible provides a further route to reduce internal forces or alter bending action to axial force action. The reverse statements can be illustrated using an example.

Figure 2.8 shows scaffolding members used to form simple trusses to support a floor above, which effectively reduces the deflection of the floor. Consequently, part of the internal forces in the floor will pass on through the truss members to the supports of the truss. This can be interpreted as converting some of the bending moments in the floor into axial forces in the truss members. Therefore, the bending moment in the floor reduces and becomes more uniform.

2.7.6 Relative Performance

The four structural concepts have been presented in the form of "The more . . . The smaller. . . ", which is obviously in a comparative sense. In other words, the four structural concepts provide an effective way to assess the relative performance of two or more similar forms of a structure for which any of the structural concepts can be used to achieve smaller deflections of the structure.

Comparing the relative performance of two forms of a structure would be more appropriate than examining their absolute performances. There are many sources for introducing errors in the analysis of structures such as inaccurate modelling. For example, connections may be neither pinned nor rigid, supports may be between fixed or pinned and material properties may not have their assumed values. As it is the model of a structure that is actually analysed rather than the structure itself, the accurate prediction of the behaviour of the structure is unlikely to be achieved in many cases. However, the relative performance of two similar structure models will remove errors from the analysis and modelling of the structures and allow a more reliable assessment of their different performances. For example, the two frames shown in Figure 2.6 involved the same degrees of error, possibly generated from the assumed pinned connections and boundary conditions and the estimated values of the elastic modulus and the cross-sectional areas of members. The calculated deflections may not be accurate but the ratio of the two calculated deflections would give a reliable indication of the relative behaviours of the two frames.

Thus, when evaluating the relative performance of two or more similar structures or forms of a structure the exact input data may not be necessary, *i.e.* the modulus of elasticity, the area and the second moment of area of a cross-section, loading and even the dimensions of the structures. This effectively simplifies the analysis while still capturing the physical essence of the problem. For example, the ratio of the deflections of the two frames in equation 2.33 is non-dimensional and the physical parameters, E, A and L, together with any other possible errors arising from assumptions made are cancelled out in the ratio validating the comparison. It is convenient and effective to analyse the relative performance of two similar structures. In the next four chapters, the relative performance of structures in pairs, one involving a structural concept and one not involving a structural concept, will be examined quantitatively to demonstrate convincingly the effect of using the structural concepts.

2.8 Implementation

The four structural concepts, interpreted intuitively from the principle of virtual work, provide a sound basis for implementation. This requires the development of physical measures to incorporate the benefits of considering structural concepts into practical cases to create more effective and efficient structures as has been shown in the previous examples in Figures 1.3, 2.6 and 2.7.

Only four structural concepts for whole structures have been discussed, but many physical measures can be developed based on these concepts. Many such physical measures are already being used in practice and there will be further measures that can be created to deal with particular cases. For example, providing a support is an effective way to lead to smaller internal forces in a structure and thus smaller deflections. Figure 2.9 illustrates four cases demonstrating different physical measures all serving for providing a support.

Figure 2.9a shows a steel prop used to support the deck of a footbridge. Part of the bridge loads is transmitted through the compression in the prop to its foundation. The prop effectively reduces the internal forces and the vertical deflections of the footbridge. As the flexural stiffness of the prop is not concerned and its axial deformation are negligible, it can be considered as a roller support to the bridge deck.

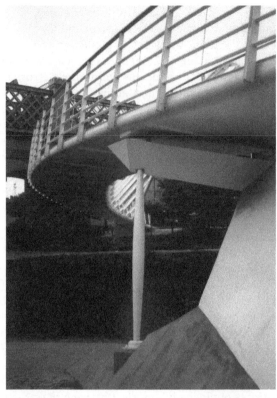

(a)

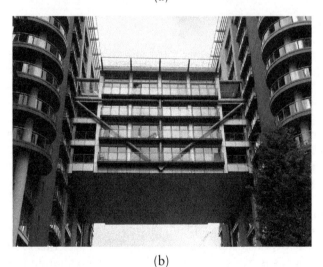

(b)

Figure 2.9 Examples of providing a support. (a) Providing a prop. (b) Providing two
         inverted triangular trusses to form vertical supports at the centre of a
         linking structure. (c) Providing tendons and a wooden bar to form a hori-
         zontal support (Courtesy of Mr Jiachen Guo, Beijing Jiaotong University,
         China). (d) Providing tendons to form elastic restraints to columns.

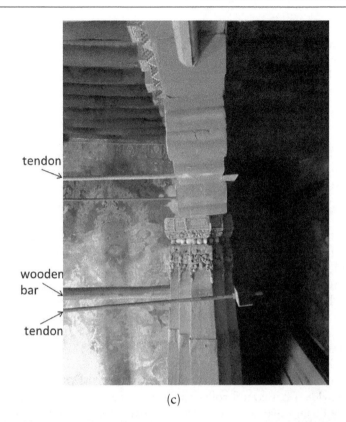

(c)

(d)

Figure 2.9 (Continued)

Figure 2.9b shows a suspended four-storey link structure between two adjacent buildings. Vertical supports are provided to the bottom centre of the link structure by two inverted triangular trusses (one each side of the link structure). The horizontal components of the tension forces in the two inclined members are balanced by the compression in the horizontal member at the height between the third and fourth storeys of the link structure and the vertical components of the tension forces are transmitted to the two adjacent buildings. The use of the truss structure effectively provides a vertical support at the central position of the link structure to achieve smaller internal forces in the link structure and thus smaller deflections. Effectively, the two inverted trusses provide elastic vertical supports to the link structure.

Figure 2.9c shows a remedial measure to provide an equivalent roller support in the lateral direction in a historic building. It can be observed that the upper wooden part of the structure has moved to its right from its supporting profiled wooden column. To prevent further horizontal movement, which might lead to at least a local collapse, physical measures were taken. Steel tendons have been attached to the upper part of the structure to limit further movement between the upper and lower parts of the structure. When limiting further movement, the steel tendon forces would be transmitted through the friction between the upper part of the structure and the lower wooden column to the lower column and then to the column support. The horizontal tendon forces on the upper part of the structure tend to pull on the lower column deforming to its left through the friction force. A further measure was developed to provide a horizontal roller support to the column. This support was implemented using two sub-measures: 1) a pair of steel tendons were placed around the lower profiled wooden columns (one of them can be seen in Figure 2.9c, which would prevent the wooden column from deforming to its right. However, the tendons only carry tension forces and a larger tension force could cause the column to deform too much to its left. In other words, the action of the tendons was in fact different to a roller support; 2) to compensate for this effect, a thicker wooden bar was placed between two lower columns to provide a force in the opposite direction to that of the force in the tendon. The combined action of the tendons and the wooden bar is like that of a roller support in the lateral direction.

Figure 2.9d shows two sets of perpendicular horizontal tendons provided at the upper ends of columns in the Palace of the Grand Master of the Knights of Rhodes. The other ends of the tendons were fixed through the walls of the room. Tension applied in the tendons in opposite directions to the columns effectively provides restraints to the columns making them more stable and compensating for aging effects. The two pairs of tendon in the two perpendicular directions act as roller supports in the two horizontal directions. This physical measure to provide additional supports to the columns is simple and effective without affecting the use of the room.

In summary, the four cases show the different implementations of a roller or spring support into structures: a prop as a rigid support, an inverted triangular truss serving as an elastic support at the bottom of a link structure between two adjacent buildings, a combination of the tension in steel tendons and the compression of a wooden bar acting as a pinned support, and tension forces applied in opposite directions serving as an elastic support. There are also many other forms of implementation measures that can be used to realise a roller or spring support to reduce internal forces and hence deflections to suit different structural situations.

2.9 Summary

Deflections and internal forces of structures are functions of applied loads that have many variations and different combinations in design. This leads to the difficulties to consider the general characteristics between deflections and internal forces of structures. In this chapter, the loading is simplified into a unit load applied on the critical point of a structure, which represents the worst loading scenario that all loads are lumped to the critical point and normalised to a unit. This avoids the investigation of the particular effects of actual loading on structures and allows revealing the general and qualitative relationships between smaller deflections and desirable distributions of internal forces of structures.

Four structural concepts have been directly and intuitively interpreted based on the principle of virtual work. These structural concepts are simple and general, and this helps their applications at least to truss and frame types of structure. Due to their simplicity and effectiveness, it is hoped that they can be used widely in practice as rules of thumb. Each of the four structural concepts has its own emphasis and characteristics and these will be discussed in the next four chapters.

Due to the interchangeability between the four structural concepts, one application can be seen as an implementation of more than one of the four structural concepts. As an example, the case in Figure 2.9b can be further examined. The provision of the inverted triangular trusses can be seen as the implementation of the fourth structural concept as part of the bending moments of the link structure is converted into the axial forces in the members of the trusses. Alternatively, it can be seen to be the realisation of the second structural concept in which the bending moments in the linking structure become smaller due to the upward force provided by the inclined members of the trusses. Therefore, the focus in the next four chapters will be on the creative use of the structural concepts rather than on exact classification of applications.

References

2.1   Gere, J. M. and Timoshenko, S. P. *Mechanics of Materials*, PWS-KENT Publishing Company, 1990, ISBN:0-534-92174-4.

2.2   Graig, R. R. *Mechanics of Materials*, John Wiley & Sons, USA, 1996.

2.3   Timoshenko, S. P. *History of Strength of Materials*, New York: McGraw-Hill Book Co., 1953.

2.4   Ji, T. Concepts for Designing Stiffer Structures, *The Structural Engineer*, 81(21), 36–42, 2013.

2.5   Yu, X. *Improving the Efficiency of Structures Using Structural Concepts*, PhD Thesis, The University of Manchester, 2012.

2.6   Ji, T., Bell, A. J. and Ellis, B. R. *Understanding and Using Structural Concepts*, Second Edition, Taylor & Francis, USA, 2016.

2.7   Huang, X. and Xie, Y. M. A Further Review of ESO Type Methods for Topology Optimisation, *Structural and Multidisciplinary Optimisation*, 41, 671–683, 2010.

Chapter 3

More Direct Internal Force Paths

3.1 Routes to Implementation

The appropriate use of bracing systems in structures is an effective way to create more direct internal force paths. Bracing systems are normally used for stabilising structures, transmitting loads and increasing lateral structural stiffness. They are ideal for use in types of structure that are sensitive to lateral loads, such as tall buildings, temporary grandstands and scaffolding structures.

Bracing systems provide direct structural expressions of internal force paths or load flow how lateral loads are transmitted through structures to their foundations. There are many, almost unlimited, options to arrange bracing members and there are large numbers of possible bracing patterns, as evidenced in existing structures. What is the most effective way to design bracing patterns?

An effective way is to follow the structural concept, *the more direct the internal force paths, the smaller the deflection*. For the purpose of application, four criteria have been intuitively developed based on this structural concept aiming to transmit a load at the critical point to the supports of a structure more directly [3.1, 3.2]:

Criterion 1: *Bracing members should be provided in each storey from the support (base) to the top of the structure;*

Criterion 2: *Bracing members in different storeys should be directly connected;*

Criterion 3: *Bracing members should be linked in a straight line where possible;*

Criterion 4: *Bracing members in the top storey and in adjacent bays should be directly connected where possible.* (This criterion is suitable to the structures that the number of bays in the horizontal direction is larger than the number of storeys in the vertical direction.)

The first criterion is obvious since the critical point for a multi-storey structure is at the top of the structure and the load at the top must be transmitted to the supports of the structure. Therefore, bracing members should be arranged in every storey of a structure. If bracing members are missing in one of the storeys, it means that the internal force path is cut off and the force has to flow along an alternative path to reach the support. In other words, the internal forces have to pass along a longer or less effective way to the supports. Consequently, the structure is likely to experience larger deflections.

There are a number of possibilities for achieving the first criterion, but the second and the third criteria suggest a way for using a *more direct force path*. Once the bracing members are directly linked, the internal forces can flow directly through them; once the bracing members are linked in a straight line, the internal forces can flow through them even more directly.

The first three criteria concern mainly the bracing arrangements in different storeys of a structure and are suitable for tall buildings for which the number of storeys is larger than the number of bays. For other types of structures, such as temporary grandstands, the number of bays is usually larger than the number of the storeys. To create shorter internal force paths or more zero-force members in such structures, the fourth criterion gives a means for considering the relationship of bracing members across the bays of the structure.

Bracing members can also be used to create alternative, and sometimes longer, internal force paths to help meet functional requirements of a structure and solve challenging technical problems.

3.2 Hand Calculation Examples

3.2.1 Effect of the Four Bracing Criteria

This example examines the effectiveness and efficiency of each of the four bracing criteria for reducing internal forces and lateral deflections of simple frames.

In order to examine the effectiveness of the four criteria for arranging bracing members, four pin-jointed plane frames are created. Each consists of four bays and two storeys and uses four bracing members. There are a total 22 members including 4 bracing members in each of the first four frames. The bracing members in the four frames are arranged in such a way that the effectiveness of each criterion given in Section 3.1 can be identified, which are shown in Figure 3.1 and their features can be summarised as follows:

> Frame A: The bracing members are arranged to satisfy the first criterion.
> Frame B: The bracing members are arranged to satisfy the first two criteria.
> Frame C: The bracing members are arranged to satisfy the first three criteria.
> Frame D: The bracing members are arranged to satisfy the four criteria.

In order to examine the effect of the bracing members that are not arranged fully following the four criteria, Frame E is created as follows:

> Frame E: Two additional bracing members are added to Frame C between the first levels and ground, which don not follow any of the four criteria.

All frame members have the same elastic modulus, *E*, and cross-sectional area, *A* with *EA = 1000*kN. The vertical and horizontal members have the same

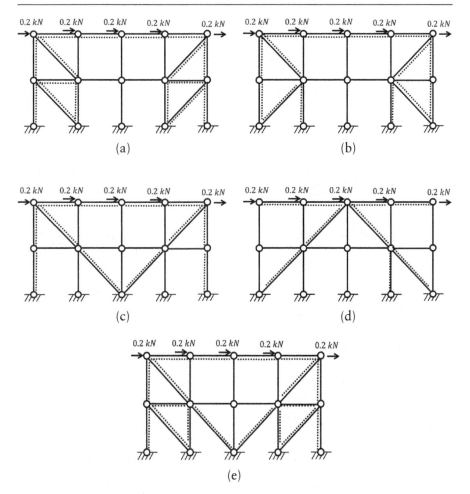

Figure 3.1 Five frames with different bracing arrangements and internal force paths
(dashed lines). (a) Frame A with six zero-force members. (b) Frame B
with eight zero-force members. (c) Frame C with ten zero-force members. (d) Frame D with 14 zero-force members. (e) Frame E with six
zero-force members.

length of $L = 1000$mm. A concentrated horizontal load of 0.2kN is applied to
each of the five top nodes of the frames. Calculate the internal forces and the
averaged deflections of the top five nodes of the five frames in the horizontal
direction.

Before determining lateral deflections and internal forces, it is possible to
identify intuitively which members are in a zero-force state. Dashed lines are
drawn next to the members that are not in a zero-force state as shown in
Figure 3.1, which indicate the internal force paths transmitting the applied
loads to the supports of the frames. The fewer the dashed lines means the
more direct the internal force paths and consequently the smaller lateral
deflection.

The five frames are statically indeterminate structures and beyond simple hand calculations. However, using the structural concept of symmetry that *symmetric structures subjected to anti-symmetric loads will lead to anti-symmetric responses*, the two vertical members on the centre lines of Frames A-E must be in a zero-force state and thus can be removed from the frames for analysis and the nodes on the centre lines have no vertical movements and can be represented by roller supports. Consequently, only halves of the five frames need to be analysed, and the first four halved frames are statically determinate, suitable for hand calculations, but the half of Frame E remains statically indeterminate and is analysed using computer software. Figure 3.2 shows the halves of the five frames equivalent in which the calculated internal forces are indicated in kN next to their members.

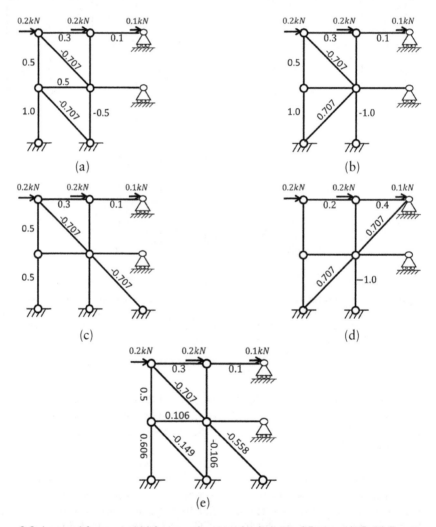

Figure 3.2 Internal forces in kN for members in the halves of Frames A-E. (a) Frame A equivalent. (b) Frame B equivalent. (c) Frame C equivalent. (d) Frame D equivalent. (e) Frame E equivalent.

The deflections of the five frames can be calculated using equation 2.16. Frame A in Figure 3.2 is selected as an illustration and equation 2.16 gives:

$$\Delta_{ave} = \sum \frac{N_i^2 L_i}{EA_i}$$

$$= \frac{[0.1^2 + 0.3^2 + 0.5^2 \times 3 + 1^2 + (-0.707)^2 \times 1.41 \times 2] \times 2 \times (1000N)(1000mm)}{(1000000N)}$$

$$= 6.52mm$$

Δ_{ave} is the averaged lateral deflection of the top five nodes of the frame. The value in the square brackets is doubled due to the contributions of the internal forces in the right half of the frame. An alternative expression is that on the left side of the previous equation is $0.5kN \times \Delta_{ave}$ for the half frame. The deflections of the other frames can be calculated in a similar manner.

To appreciate the effect of the four criteria for realising more direct internal force paths, Table 3.1 summarises and compares five sets of results calculated for the five frames based on Figures 3.1 and 3.2. The five sets of results in rows are:

1. The total numbers of zero-force members which can be counted directly from Figure 3.1.
2. The largest absolute values of the internal forces in the horizontal and vertical members which can be found from the internal forces indicated in Figure 3.2.
3. The averaged horizontal deflections of the five top nodes which are calculated using equation 2.16 for the unit load, which is distributed equally to the five top nodes of the frames.
4. The relative horizontal deflections normalised to that of Frame A.
5. The relative horizontal stiffnesses which are the inverses of the relative deflections in (4).

Observations from Table 3.1 and Figures 3.1 and 3.2 can be discussed further:

- *Frame A (satisfying the first criterion)*: The frame has a conventional form of bracing and the horizontal loads at the top are transmitted to the

Table 3.1 A Summary of the Results of the Five Frames (Figures 3.1 and 3.2)

	Frame	A	B	C	D	E
1	Number of zero-force members	6	8	10	14	6
2	The largest internal force (kN) in the vertical and horizontal members	1.0	1.0	0.5	0.4	0.6
3	The averaged horizontal deflection of the five top nodes (mm)	6.52	6.03	4.03	3.23	3.90
4	Relative deflection		1.0	0.925	0.618	0.495	0.598
5	Relative stiffness		1.0	1.08	1.62	2.02	1.68

supports through the bracing, vertical and horizontal members. The internal force paths can be examined more closely (Figure 3.2a). The loads pass through the side vertical members and the bracing members in the upper storey and the internal force in the bracing member then passes to the connected vertical and horizontal members at the top of the lower storey. The internal force in the horizontal member passes to the bracing and vertical member in the lower storey and then to the supports. The internal forces in the side vertical members are generated to balance the vertical components of the internal forces of the bracing members. This relatively long internal force path leaves only two members with zero force, *i.e.* a total of six zero-force members in the full frame.

• *Frame B (satisfying the first two criteria)*: It can be seen from Figure 3.2b that the internal force in the bracing member in the upper storey passes directly to the bracing and vertical members in the lower storey without passing through the horizontal member at the top of the lower storey. Frame B provides a shorter force path than Frame A with one more zero-force member in the equivalent half frame and thus has a smaller deflection than Frame A.

• *Frame C (satisfying the first three criteria)*: Figure 3.2c shows that a more direct force path is created with one vertical member in the lower storey, which has the largest force in Frame B, becoming zero-force member. The shorter force path produces an even smaller deflection, as expected. The third criterion is particularly efficient for not only creating a more direct force path but also for removing the largest internal force, which effectively reduces the deflection in comparison with that of Frames A and B.

• *Frame D (satisfying all four criteria):* In Frame C, to transmit the lateral loads at the top nodes, where bracing members are involved, forces in vertical members have to be generated to balance the vertical components of the forces in the bracing members (Figure 3.2c). In Frame D two bracing members with mirror orientations are connected at the top central node, with one member in compression and the other in tension. From Figures 3.1d and 3.2d it can be seen that the horizontal components of the forces in these bracing members balance the external lateral loads while the vertical components of the forces are self-balancing. Therefore, all vertical members are in a zero-force state and Frame D leads to the lowest deflection of Frames A to D.

• *Frame E (satisfying the first three criteria and having two additional bracing members that do not follow any criteria)*: Two more members are added to Frame C to form Frame E, but comparison between Frames D and E indicates that bracing members which follow the criteria set out can lead to a smaller deflection than providing more bracing members which do not fully follow the criteria. As Frame E has two added members compared to Frame C, it should be stiffer than Frame C as expected. In comparison with Frame D, in Frame E one bracing member has a smaller force of 0.558kN against 0.707kN while five more members are in a forced state with a maximum force value of 0.606kN. Therefore, Frame E, which uses more bracing members than Frame D, has a larger deflection than frame D.

It can also be observed from Table 3.1 that the structure has a smaller deflection and is stiffer when the internal forces are smaller and more uniformly distributed although the first four criteria are derived on the basis of the structural concept of more direct force paths. These examples are simple, and the variation of bracing arrangements is limited, but they do demonstrate the effectiveness and efficiency of the criteria that are based on the structural concept of more direct internal force paths.

3.2.2 The Most and Least Effective Bracing Patterns for a Simple Frame

This example identifies the most and least effective bracing patterns through examining many thousands of bracing patterns of a four-bay and four-storey frame.

Figure 3.3 shows a four-bay and four-storey pin-jointed frame structure that is composed of sixteen horizontal and twenty vertical bar members. In this example a panel is defined as the empty area enclosed by two horizontal and two vertical members. Therefore, the frame in Figure 3.3a has 16 panels. The frame is stabilised and stiffened using eight bracing members following the rules: 1) at each storey, two of the four panels are braced; 2) in each of the braced panels, there are two possible bracing orientations (Figure 3.3b) [3.3].

Selecting any two from the four panels in a storey gives six options for bracing in each storey (Figure 3.3b), *i.e.* $\dfrac{4!}{2!(4-2)!}$. For each braced panel, there are two possible bracing orientations. Thus, there are 24 bracing options in each storey, *i.e.* 6 x 2 x 2 = 24. For all four storeys, a total of 24^4 =331776 patterns are possible. When only symmetric bracing arrangements are considered, the number of possible bracing patterns reduces to 256 cases, *i.e.* $(2 \times 2)^4 = 256$.

To simplify the analysis, it is considered that the vertical and horizontal members have the same length, $b = a = 1000mm$, all members have the same cross-sectional area A and elastic modulus E, and EA = 1000kN. A pair of horizontal forces, each with a value of 0.5kN, are applied anti-symmetrically at the two top corner nodes of the frame. The lateral stiffness of the frame is defined as the inverse of the average of the lateral deflections of the two nodes. The maximum horizontal deflections and internal forces of the frame with different bracing patterns will be compared.

The ANSYS finite element method package was used to calculate the maximum lateral deflections of all 331776 cases and all 256 symmetric cases. The maximum horizontal deflections for all 331776 cases are ranked from the smallest to the largest and demonstrate that the X braced frame, shown

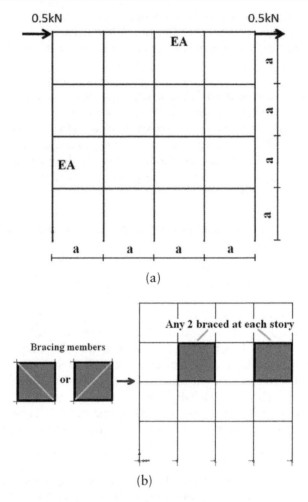

Figure 3.3 A four-bay and four-storey frame: a) Geometry of the frame; (b) Any two braced panels in each storey and bracing orientations.

in Figure 1.3 and Figure 3.5a, has the smallest deflection for the four-bay and four-storey square paneled frame with all members being the same cross-section.

When a symmetric structure is subjected to anti-symmetric loads, the internal forces and deformations of the structure must be anti-symmetric. Hence the axial forces in the central vertical members of the frames must be zero and the nodes in the central lines have no vertical displacements. Thus, each of the frames can be equivalently simplified into a half frame, which becomes a statically determinate structure, as shown in Figure 3.4b. This greatly simplifies the analysis of the symmetrically braced frames and allows a hand calculation to

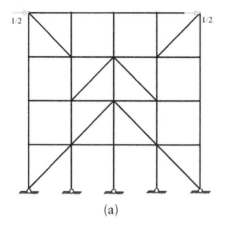
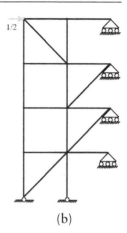

(a)

(b)

Figure 3.4 Using symmetry to simplify the frame model: (a) A whole structure; (b) A half equivalent structure.

be conducted for checking and for gaining an insight into the behaviour of the braced frames.

Figures 3.5 and 3.6 present the six most effective and six least effective bracing patterns defined by the smallest and largest horizontal deflections from 256 symmetric bracing patterns, in which the internal forces are indicated next to the members and the maximum deflections are given at the bottoms of corresponding frames.

The bracing patterns in each of the two groups of frames are similar with small variations. However, the differences in the bracing patterns between the two groups are obvious and can be summarised as follows:

- The six bracing patterns with the smallest lateral deflections have diagonally braced panels in general and at least two bracing members are linked in straight lines (Figure 3.5).
- The six bracing patterns with the largest lateral deflections have two independent vertically braced panels and the bracing members are mainly placed in parallel to each other (Figure 3.6).

These observations suggest that frames should be braced diagonally across the whole width of a structure and bracing members should be linked in a straight line where possible. Alternatively, it should be avoided that bracing members are arranged in independent vertical panels and are placed in parallel.

Figure 3.7 shows a pair of physical models that resemble the frame in Figures 3.6e and the frame in Figure 3.5a for which detailed hand calculations are given in Section 1.2. The maximum lateral deflections of the two frames are 29.16mm and 7.65mm respectively, which gives a stiffness ratio of 29.16/7.656 = 3.81 for the two frames. With such a large difference in stiffness, it is easy to feel the relative stiffness of the two frame models by pushing the top left corners horizontally.

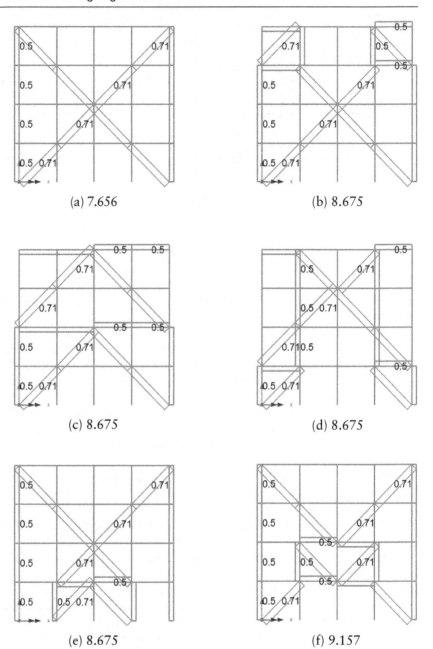

Figure 3.5 The six most effective bracing patterns, the internal forces (kN) and the maximum deflections (mm).

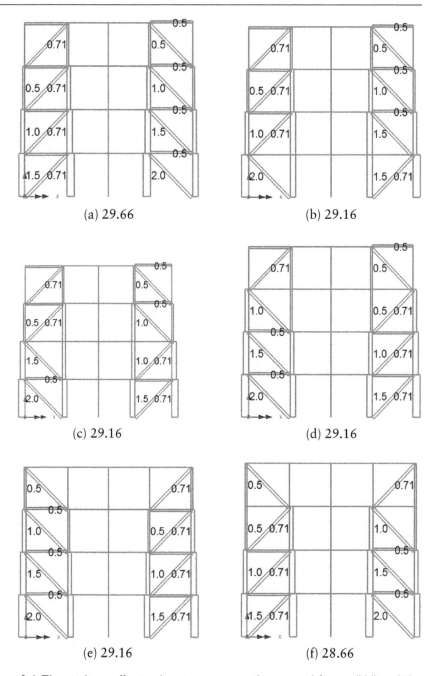

Figure 3.6 The six least effective bracing patterns, the internal forces (kN) and the maximum deflections (mm).

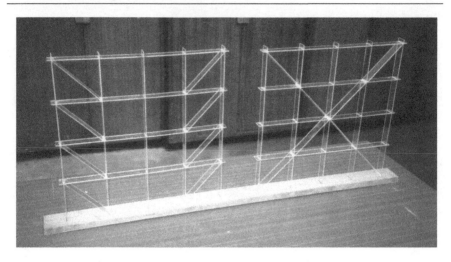

Figure 3.7 Physical models used to feel their relative stiffness [3.4].

The 46 stiffest and the 50 least stiff frames among the 256 symmetrically braced cases can also be analysed by hand using equation 2.16. The 46 stiffest frames are selected because the 46th to the 60th frames have the same stiffness. The deflections of the frames can be classified into three groups, *i.e.* these contributed by the horizontal members (H for δ_H), the vertical members (V for δ_V) and the diagonal members (D for δ_D). Equation 2.16 can be written for each of the 96 cases as [3.3]:

$$\delta = \delta_H + \delta_V + \delta_D = \sum_{i=1}^{44}\left(\frac{N_i^2 L_i}{EA}\right) = \left(\sum_{Hi=1}^{16} N_{Hi}^2 + \sum_{Vi=1}^{20} N_{Vi}^2 + \sqrt{2}\sum_{Di=1}^{8} N_{Di}^2\right)\frac{a}{EA} \quad (3.1)$$

The deflections, δ_H, δ_V, δ_D and δ, for the 96 cases have been calculated and are presented graphically in Figures 3.8a and 3.8b for the 46 stiffest cases and the 50 least stiff cases respectively. It can be observed from Figure 3.8 that:

1. The lateral deflections contributed by the diagonal members (δ_D) are constant for all cases.
2. The lateral deflections contributed by the horizontal members (δ_H) are approximately constant for all cases and are smaller than those of the diagonal members.
3. For the 46 stiffest cases, the lateral deflections contributed by the vertical members (δ_V) vary insignificantly and are smaller than those of the diagonal members.
4. For the 50 least stiff cases, the lateral deflections contributed by the vertical members (δ_V) vary significantly and are much larger than those of the bracing members.
5. For the 46 stiffest frames (Figure 3.8a), δ_H and δ_V have similar magnitudes. For the 50 least stiff frames (Figure 3.8b), δ_V is much larger than δ_H.

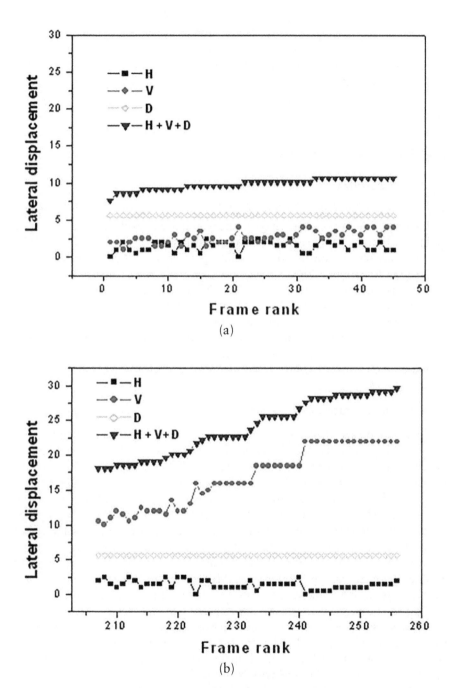

Figure 3.8 Lateral deflections contributed by vertical, horizontal and diagonal members: (a) For the 46 stiffest frames. (b) For the 50 least stiff frames.

Because δ_D is constant and δ_H varies insignificantly for the 96 cases (Figure 3.8), the deflections and the relative deflections (the rank) of the 256 cases are basically controlled by δ_V. Therefore, bracing patterns that produce large values of δ_V, or large internal forces in the vertical members, should be avoided. In other words, the four criteria provide a practical guide for designing bracing patterns to reduce values of δ_V, the lateral deflection contributed by the internal forces in the vertical members.

3.3 Practical Examples

3.3.1 Tall Buildings

There are different classifications of tall buildings. It is generally thought that multi-storey structures between 35m and 100m are considered to be high-rise buildings. Buildings higher than 100m are termed skyscrapers, buildings 300m or higher are termed super tall and buildings 600m or taller are termed mega-tall. The taller the building, the more susceptible it is to wind loads. Consequently, different structural systems have been developed to deal with height, such as rigid frame systems, shear wall systems, tube systems including framed tubes, braced tubes and tube-in-tube systems.

The bracing criteria shown in Section 3.1 will be used to examine the effectiveness and efficiency of some tall buildings in which bracing systems have been used including braced frame systems and braced tube systems.

3.3.1.1 John Hancock Center, Chicago

The John Hancock Center (or Tower), a 100-storey 344m tall building, was built in Chicago that has the nickname of "the windy city". The Tower was built in 1969 when computers were little used in building design. The form of the building (Figure 3.9) shows that the wide base of the building provides greater structural stability and the narrow upmost part effectively reduces the lateral wind forces. The structural engineer Fazlur Khan and his collaborators proposed an exterior-braced frame tube structure. Five and a half huge X bracings, each across 18 storeys, were used in each of the four sides of the building. Horizontal members were placed between the connections of the bracing members. An advance on the usual steel-framed tube, this design added global cross-bracing to the perimeter frame to increase the lateral stiffness of the structure (Figure 3.9).

It can be observed from Figure 3.9 that the external frame columns, global cross-bracings and beams form a huge exterior trussed tube which is highly effective for resisting lateral loads. The structural expression of the tube and the great effectiveness of its lateral resistance is highly harmonised. Some $15 million was saved on the conventional steelwork by using the huge cross-braces [3.5, 3.6]. It was regarded as an extremely economical design which achieved the required stiffness to make the building stable. One of the reasons for this success was that the required lateral stiffness of the structure was achieved by using the cross-braces.

The structural effectiveness and efficiency of the John Hancock Center can be explained in alternative ways. For example, "The form is especially efficient

(a)

Figure 3.9 Bracing systems used in the John Hancock Center, Chicago, satisfying the first three criteria. (a) The building (Courtesy of Mr. Nicolas Janberg, structurae.net, Germany). (b) A closer look.

in the Hancock Tower because the diagonals tie together the otherwise widely spaced columns, thus distributing the vertical forces evenly among them" [3.7]. It was not clear how Fazlur Khan and his collaborators generated the idea of using the huge cross braces, but this ingenious idea can be explained using the implementation criteria in Section 3.1. It is observed from Figure 3.9 that the global X bracing of the building ideally meets with the first three

(b)

Figure 3.9 (Continued)

criteria (bracing members from the top to the bottom of the building and bracing members linked and linked in a straight line where possible), which is an
implementation of the structural concept of more direct internal force paths.
Therefore, it may be said that the use of the huge cross braces creates more
direct load paths that led to larger lateral stiffness and hence smaller deflections of the structure when subjected to wind loads.

Similar huge global X braces can be observed in other well-known buildings. Resembling the global steel X bracing in the John Hancock Tower, global
concrete X braces were used in the 60-storey Onterie Center also in Chicago.
These global X braces were achieved by creating a series of solid window

spaces running diagonally along the exterior of the building as shown in Figure 3.10a. It can be observed from Figure 3.10a that these effective "bracing members" on one side satisfy the first three criteria, while on the adjacent side they meet with the first two criteria in Section 3.1. Based on the understanding obtained from the examples in Section 3.2.2, it can be deduced that the

(a)

Figure 3.10 The X braces without using beams. (a) Concrete bracing in the Onterie Center, Chicago (Courtesy of Mr. Nicolas Janberg, structurae.net, Germany). (b) Steel bracing in the Bank of China, Hong Kong.

(b)

Figure 3.10 (Continued)

stiffness contributed by the effective X bracing would be much larger than that of the snake-like bracing on the adjacent side of the building.

The use of a similar global X bracing pattern is observed in the Bank of China building in Hong Kong, which has also been regarded as an efficient and elegant design. The lights placed along the braces and columns seem to illuminate the internal force paths in the building as shown in Figure 3.10b.

3.3.1.2 Leadenhall Building, London

The Leadenhall Building, located in the centre of London, is a 224m tall commercial office tower. It is commonly known as the "Cheesegrater" because of the unique tilted elevation and steel diagrid structure [3.8]. The main difference between an X braced structure, such as the John Hancock Center, and the diagrid structure is that there are no columns or vertical members in diagrid structures. Figure 3.11 shows the front and side views of the building.

To maximise internal flexibility, a perimeter mega-frame structure is used to form a closed braced tube around all four sides of the building as shown

(a)

Figure 3.11 Leadenhall Building, London. (a) Front view (Courtesy of Mr. Nicolas Janberg, structurae.net, Germany). (b) Side view.

(b)

Figure 3.11 (Continued)

in Figure 3.12. It can be observed that the mega-frame has a varied geometrical pattern. The South elevation, the front view, in Figures 3.11a and 3.12a, shows a diagrid system while the frames in the other three sides of the building are effectively braced frames. The South frame consists of diagonal members spanning vertically between beams at the mega levels of 28m and horizontally between the mega nodes at 16m centres. The frames in the East and West elevations are comprised of columns spaced at 10.5m centres joined to diagonal bracing members and beams at each mega level, resulting in an asymmetrical geometry as shown in Figures 3.11b and 3.12b.

Diagrid structures are very effective for resisting lateral loads as the diagonal members provide more direct internal force paths to transmit the lateral loads,

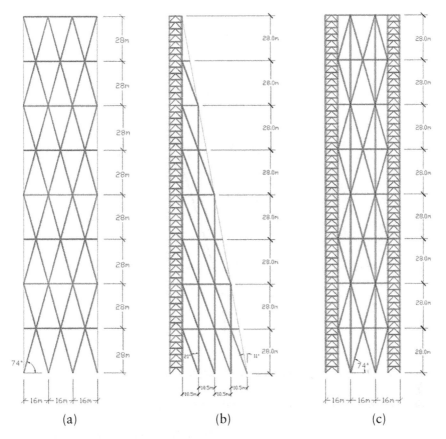

Figure 3.12 Elevation of the Leadenhall Building: (a) South Frame, (b) East/West Frame, (c) North Frame [3.9].

mainly through axial forces rather than bending moments, to the supports of the structures (the efficiency of inclined members to transmit lateral loads can be seen in the hand calculation examples in Section 6.2.2). However, they appear to be less effective for transmitting vertical loads than conventional columns. The horizontal members between the mega nodes compensate for this weakness.

Examining a typical unit subjected to vertical loads as illustrated in Figure 3.13, it can be seen that the vertical loads tend to make nodes A and B deform toward each other while nodes B and C tend to move apart from each other. However, the horizontal member, CD, in the central position ties nodes C and D to prevent them from deforming apart from each other, which in turn prevents A and B from deforming toward each other. This makes the unit much stiffer in the vertical direction. Due to the action of member CD, the vertical loads are transmitted to the supports mainly by axial forces rather than by bending moments through the inclined members.

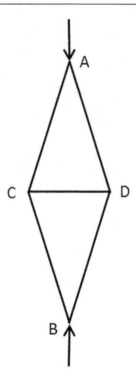

Figure 3.13 A typical diagrid unit subjected to vertical loads.

It can be seen from Figures 3.11 and 3.12 that the geometries of the Lead-enhall South and East/West frames do not include load bearing members at the corners where the two frames meet. When horizontal loads are applied in the plane of the East/West frames in Figure 3.12b, large portions of the forces are transmitted to the bracing members then flow to vertical members through turns rather than along straight lines. Considering structural efficiency alone, inclined edge members could be added to the frame which would create more direct internal force paths as shown in Figure 3.14b, leading to a stiffer struc-ture. This edge member would also serve for the South frame (Figure 3.14a) by framing the diagrid structure which would lead to smaller internal forces and a more uniform distribution of internal forces. To check this intuitive under-standing, finite element models of the South and East/West frames without and with the edge members (Figures 3.12a, 3.12b, 3.14a and 3.14b) were created for analysis. A unit concentrated load is applied at the top of the frame models. The inverse of the lateral deflection at the loading point is considered as the lateral static stiffness of the frame model. If for this example the efficiency (e) of a structure is defined as the ratio of the lateral static stiffness (K) to the total mass (M) of the frame model, the efficiencies of the South and West/East frame models have been determined as follows:

$$e = \frac{K}{M} \tag{3.2}$$

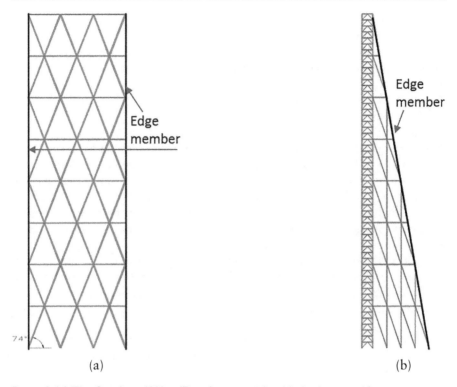

Figure 3.14 The South and West/East frames with added edge members.

The physical meaning of the efficiency is the lateral static stiffness contributed by a unit structural mass. To remove the modelling errors involved, the ratio of the efficiencies of the models with edge members (e_W) to that without the edge members (e_{WO}) is defined as:

$$R = \frac{e_W}{e_{WO}} \tag{3.3}$$

The results show that the ratio of efficiency for the South frame model is 1.72 and for the West/East frame model is 1.24, indicating that the frames with the added edge members are more efficient. This comparison examines only the efficiency of the frames without considering any other design requirements.

3.3.2 Temporary Grandstands

Temporary structures are ideal temporary solutions for temporary purposes. Temporary grandstands are frequently used at indoor and outdoor activities such as tennis tournaments, where the spectators are usually sedentary, and pop concerts, where the audiences may move energetically following music beats. Temporary grandstands are designed to be erected and demounted easily and quickly and are usually of lightweight construction with temporary supports and are therefore relatively sensitive to dynamic loads. Unlike permanent grandstands, temporary grandstands are normally supported by many vertical members so that vertical stiffnesses of the structures are not a design concern. However, temporary grandstands must possess sufficient transverse and longitudinal stiffnesses to resist horizontal loads induced by wind and by spectators' movements [3.10]. Bracing members are normally used to stiffen temporary grandstands.

The structural safety of temporary grandstands had been considered to be an important issue following several incidents, the most serious being the collapse of the rear part of a temporary grandstand in Corsica in May 1992. Subsequently, the Building Research Establishment, UK, tested 50 demountable stands of fifteen different types over several years [3.11]. The seating capacities of the grandstands varied from 243 to 3500. Only one stand had a vertical natural frequency below 8.4 Hz (at 7.9 Hz), indicating that there was no concern for human induced vibration in the vertical direction. However, the natural frequencies in the two horizontal directions were low. Table 3.2 summarises the distribution of the natural frequencies in the sway and front-to-back directions.

The relatively low natural frequencies indicated that the structures had relatively low stiffnesses in the horizontal directions. The structural characteristics of temporary grandstands can be observed from the many structures tested:

1. They were normally assembled using slender circular steel tubes, usually using the same cross-section with a small second moment of area, and the links between the vertical and horizontal members were closer to pinned connections than to rigid connections. Therefore, the frames which were formed from horizontal and vertical members had very low lateral stiffnesses as limited frame action could be developed.
2. The vertical members of these grandstands were footed directly onto the surface of the ground. Such footing conditions are regarded as pinned supports.
3. Temporary grandstands had different sizes and heights.
4. Bracing members were provided in most of the structures with many variations of bracing patterns.

The first two observations are common for most temporary grandstands and are not the main factors responsible for the low natural frequencies in the two horizontal directions which tend to be even lower in taller temporary grandstands. An intuitive understanding of the low natural frequencies (or

Table 3.2 Principal Horizontal Natural Frequencies of Temporary Grandstands [3.10]

Frequency (Hz)	Number of stands	
	Sway (longitudinal)	Front-to-back
Under 3.0	15	10
3.0–3.9	17	13
4.0–4.9	13	9
5.0 or over	5	18

stiffnesses) in the horizontal directions were that ineffective bracing arrangements were used. These site experiments and observations had generated the study on effective bracing systems for temporary grandstands and the development of the criteria for arranging bracing members [3.1].

3.3.2.1 Collapse of a Temporary Grandstand in Corsica, France

On 5 May 1992 a temporary grandstand at Furiani Stadium in Bastia, Corsica, France, collapsed, killing 18 people and injuring 2300. On that day SC Bastia faced Olympique de Marseille for a semi-final football match in the French Cup that was the premier knockout cup competition in French football organised by the French Football Federation (FFF). In order to accommodate a large-capacity crowd, an additional temporary grandstand was erected at the back of an existing grandstand to increase the number of seats by 50%. Local authorities approved the project without restrictions. This added rear part of the grandstand collapsed at 20:20 shortly before the scheduled start of the match. An investigation of the disaster concluded that there had been several violations of rules concerning the construction of the temporary grandstand. Problems were also identified in the management of ticketing and in the attitudes of sporting and municipal executives.

Figure 3.15 shows the cross-section of the grandstand that had the front part and the back part in the north-south direction. The front part consisted of six 3m bays making a total width of 18m and the back part had four 3m bays with a total width of 12m. The maximum height of stand was about 11m. The internal force paths of the temporary grandstand (back part) can now be examined using the bracing criteria based on the information in Figure 3.15.

1. There were no bracing members in Bay 9, and the two horizontal members and the seating unit in the bay simply linked bays 8 and 10. This did not

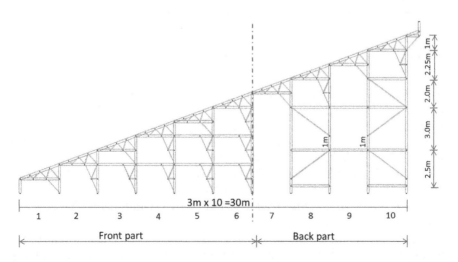

Figure 3.15 Cross-section of the collapsed temporary grandstand in Corsica, France.

contribute the lateral stiffness of the bay to the back part of the grand-stand. If there had been no links between the seating decks in bays 8 and 10, the lateral stiffness of the rear part of the grandstand would have been the sum of the stiffnesses of the two independent bays, 8 and 10, rather than that of the whole of the back part of the grandstand.

2.   There was a weak connection at a point between the front and back parts of the stand, *i.e.* the connection between bays 6 and 7. It appeared that the front part was assembled using standard units and was much stiffer than the back part. However, advantage was not taken of this leaving the taller back part of the stand with insufficient supports. If bracing members had been arranged in bay 7 to unify the two parts of the grandstand, the back part would have been much stiffer and stronger, and the collapse might have been avoided.

3.   There were large eccentricities (1.0m) between the ends of the two middle bracing members and the intersection points of the horizontal and verti-cal members in bays 8 and 10. The internal forces in the middle bracing members were transmitted to the vertical members to which they were connected and the slender vertical members had to bend to transmit the eccentric forces to the intersection points.

Through these observations and their qualitative interpretation, it can be real-ised that the back part of the grandstand had a low natural frequency and stiffness in the lateral direction and was thus susceptible to human induced dynamic loads. The identification of the weakness effectively suggests ways to improve the grandstand to have much higher stiffness in the lateral direction:

1.   Providing bracing members in bay 9 to allow bays 8, 9 and 10 working as a whole.

2.   Providing bracing and horizontal members in bay 7 to allow the front and back parts of the grandstand working as a whole.

3.   Placing bracing members at the connections of vertical and horizontal members to allow for more direct internal force paths in bays 8 and 10.

3.3.2.2 A Temporary Grandstand in Eastbourne, UK

Figure 3.16 shows the back and side views of a temporary grandstand erected for the International Women's Tennis Championship in June 1992 in East-bourne, UK. The stand consisted of 38 trusses constructed and slotted together using a specially made scaffolding system. Eight vertical members carried each truss to adjustable supports on wooden bases. The grandstand could hold just under 2700 people accommodated in 28 rows of seats, with up to 100 seats in each row. The stand was about 23.2m from front to back, rising from 2.5m at the front to 10.6m at the back. Its length was estimated to be about 60m.

The bracing patterns in the back and side of the grandstand can be identified from Figure 3.16. The back frame of the grandstand, with 25 bays, is shown in Figure 3.17a, in which alternative bays were braced from the bottom to the top. A bracing member was placed at the first storey level in all the other bays. Among the 50 temporary grandstands tested, this had the best bracing [3.10].

(a)

(b)

Figure 3.16 A temporary grandstand in Eastbourne, UK. (a) The back view. (b) The side view.

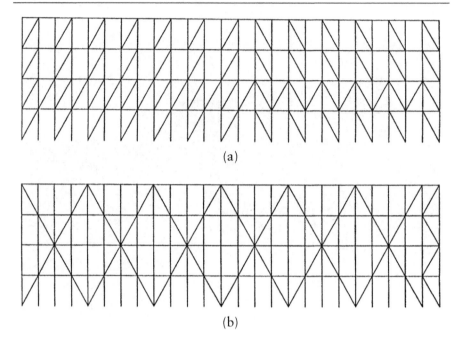

(a)

(b)

Figure 3.17 Design of bracing patterns. (a) Frame A based on the design shown in
Figure 3.16a. (b) Frame B based on the four bracing criteria.

The vibration of the stand in the sway (longitudinal) direction was however a major concern. Vibration tests showed that the fundamental natural
frequency of the stand in this direction changed from about 2.7 Hz when the
stand was empty to about 1.7 Hz when the stand was fairly full.

It can be seen from Figure 3.17a that the bracing members were arranged
from the bottom to the top of the grandstand which satisfies Criterion 1 and
some of the bracing members were linked in straight lines across the first three
storeys. However, the linkage in a straight line did not pass throughout the
full height of the grandstand and no bracing members meet at the tops of the
structure. To fulfil all four criteria, the bracing pattern can be redesigned as
shown in Figure 3.17b. The redesign is straightforward if the four criteria are
implemented without considering anything else, such as safety, economy and
elegance of the structure. To compare the effectiveness and efficiency of the
two bracing patterns for the grandstand, computer analyses were conducted.
Table 3.3 compares the static stiffnesses, the fundamental natural frequencies
and the numbers of bracing members used of the two braced frames.

The comparison in Table 3.3 shows that the lateral stiffness of Frame B with
the improved bracing pattern is much larger, being 284% of that of Frame
A with the original bracing pattern. The ratio of the fundamental natural
frequencies is 169% as the stiffness is proportional to the natural frequency
squared. The significant increase of the stiffness is due to the improved bracing pattern providing far more direct internal force paths, as described in

Table 3.3 Comparison of the Lateral Stiffnesses and Natural Frequencies of Frames
A and B

	Lateral stiffness	Fundamental natural frequency	Number of bracing members
Frame A: with the original bracing pattern (Figure 3.17a)	3.16 MN/m	1.96 Hz	64
Frame B: with the improved bracing pattern (Figure 3.17b)	8.96 MN/m	3.31 Hz	52
Ratio (Model B/Model A)	2.84	1.69	0.81

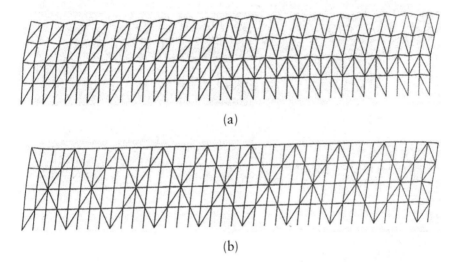

(a)

(b)

Figure 3.18 Comparison of the shapes of the fundamental modes of the two frames.
(a) Frame A. (b) Frame B.

Sections 3.1 and 3.2. In terms of the consumption of material, Frame B uses
19% less of bracing members than Frame A.

Figure 3.18 compares the shapes of the fundamental modes of vibration of
the two frames in which the maximum movements are normalised to the same
value. It shows that Frame B with the improved bracing pattern displays global
deformations while Frame A with the original bracing pattern displays both
global deformations and local deformations at the top nodes, which further
demonstrates the effectiveness and efficiency of the improved bracing pattern.

Table 3.3 shows that Frame B, with the improved bracing pattern based
on the four bracing criteria, is more effective (possessing much larger stiffness to reduce deflections), more efficient (using a smaller number of bracing
members). When examining the appearance of the two frames in Figure 3.17,
it seems that Frame B, following the bracing criteria, is more elegant than the
Frame A, with the original bracing pattern.

3.3.2.3 Two Further Cases

Figure 3.19a shows a temporary grandstand used for the British Grand Prix in Silverstone, UK. It can be observed from the back of the stand that no bracing members were provided which led to a low stiffness and a low fundamental natural frequency in the lateral direction. Fortunately, the spectators watching

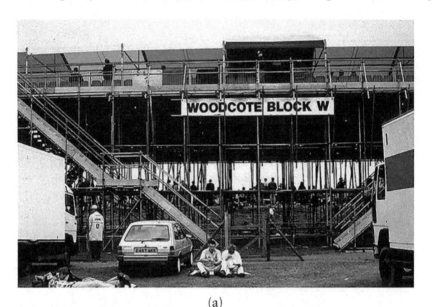

(a)

(b)

Figure 3.19 Bracing patterns of two temporary grandstands. (a) Without using bracing members. (b) Assembled using standard units.

the motor racing were sedentary and the temporary grandstand survived its use for motor racing events. Such a temporary grandstand, however, could not be used for pop concerts or for football events in which human dynamic loads would be experienced, as resonance might occur in either the lateral or the front-to-back direction of the temporary grandstand.

Figure 3.19b shows a temporary grandstand that was assembled using standard units in which the units were only connected by the relatively heavy and stiff seating decks at the top of the grandstand. The advantage of using this type of grandstand is that it is quick and easy to erect. However, the drawback is that it has low lateral natural frequencies. For an easy understanding, the temporary grandstand is resembled as a simple plane model as shown in Figure 3.20. The plane model consisting of four equally spaced plane units that are linked at their tops through a rigid plate. If a new unit is added to the model, the mass on two bays will also be added. Consider that each unit has a lateral stiffness of k and a lumped mass of m at its top, the lateral stiffness of the plane model is $4k$ and the mass at the top of the model is $(2 \times 4 - 1)m = 7m$, i.e. the sum of the stiffness of the four individual units and the sum of the mass of the seven bays. If the temporary grandstand (Figure 3.19b) consists of n units and each unit has its lateral stiffness of k and a deck mass of m, the lateral stiffness of the stand would be nk and the total mass on the top would be $(2n-1)m$. The natural frequency of the whole grandstand in the lateral direction (f_w) would be close to that of a typical unit with the mass for two bays (f_u), i.e.:

$$f_w = \frac{1}{2\pi} \sqrt{\frac{nk}{(2n-1)m}} \approx \frac{1}{2\pi} \sqrt{\frac{k}{2m}} = f_u \tag{3.4}$$

Equation 3.4 indicates that the temporary grandstand takes more units will not increase its lateral natural frequency as the mass of the grandstand increases proportionally to the stiffness.

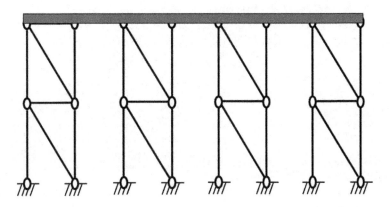

Figure 3.20 A plane model to resemble the grandstand in Figure 3.19b.

3.3.3 Scaffolding Structures

Scaffolding structures are temporary structures that are used primarily to provide temporary access during the construction or renovation of buildings and other structures. The design of scaffolding imposes some restrictions that can be ignored in the design of other structures. For example, scaffolding structures must be easily assembled and taken apart and the components should also be relatively light to permit manhandling. Scaffolding structures are often erected using simple units and slender members and the connections, bracing patterns and load paths are not always designed appropriately. Many projects require very large scaffolding structures which must possess sufficient lateral stiffness to ensure that all the loads acting on them can be transmitted safely to their supports. Although scaffolding structures are light and temporary, their design should be taken seriously. The concept of direct force paths and the four criteria are applicable to scaffolding structures.

3.3.3.1 Collapse of a Scaffolding Structure, Manchester

The scaffolding structure shown in Figure 3.21 collapsed in 1993 [3.12], though no specific explanation was given. Using the structural concept of direct internal force paths and the understanding gained from the examples in Section 3.2, the cause of the incident may be suggested. No diagonal (bracing) members were provided in the scaffolding structure, *i.e.* no direct internal force paths were provided. The scaffolding structure worked as an unbraced frame structure and the lateral loads, such as wind loads, on the structure had to be transmitted to its supports through bending of the slender vertical scaffolding members. The structure did not have enough lateral stiffness and collapsed under wind loads only.

Figure 3.21 Collapse of a scaffolding structure (Courtesy of Mr John Anderson).

3.3.3.2 Lack of Direct Internal Force Paths

For the convenience of erection of the scaffolding structures, standard proprietary units were used. The unit used in the scaffolding structure shown in Figure 3.22a consisted of two horizontal members and two short diagonal bracing members supported by two vertical members. The unit is useful for transmitting vertical loads applied to the top horizontal member to the vertical members. It is equivalent to a thick beam in the overall scaffolding structure, which is effectively a deep beam and slender column structure. The diagonal members used in the structure do not provide the force paths to transmit the lateral loads on the structure from the top to the bottom of the structure and do not follow the basic criteria for arranging bracing members. Therefore, the scaffolding structure can be judged to have low lateral stiffness.

Figure 3.22b shows another example where the scaffolding structure works as a frame structure with strong beams, the trusses, and weak columns. It would be inconvenient to place vertical members in the entrance area and the lower truss over the entrance supports two vertical members above. The scaffolding structure resists lateral loads mainly through bending rather than through axial forces in the vertical members which provides less effective lateral resistance. In addition, the slender vertical members are not ideal for transmitting bending. Judging by the four bracing criteria, this scaffolding structure has a lack of internal force paths from the top to the bottom of the structure, and no direct internal force paths have been created. It can be concluded that the scaffolding structure has a low lateral stiffness.

3.4 Further Comments

The structural concept of more direct internal force paths has been implemented by using appropriate bracing patterns that can be applied to tall buildings, temporary grandstands and scaffolding structures. The hand calculation examples, and the practical cases, demonstrate that the use of the structural concept or the implementation criteria can make structures stiffer (experiencing smaller deformation), more efficient and perhaps more elegant. There are other implementation measures to be explored and these may be observed from existing structures or developed from the structural concept itself.

Internal force paths or load paths can be designed to solve other practical and challenging structural problems as can be observed from practical examples. One such example is seen at the entrance of the Cannon Street Underground Station in London, the upper eight storeys of the building cantilever a distance from the building supports as shown in Figure 3.23. How do the loads of the cantilever building transmit to the supports of the building? To understand the load paths, a simple diagram may be drawn for qualitative analysis which aims to capture the physical essence of the load paths but omits some less important details. Figure 3.24 shows a model of the side façade of the building based on the photos in Figure 3.23 and this acts like a truss structure.

(a)

(b)

Figure 3.22 Scaffolding structures assembled from proprietary units but lacking direct internal force paths.

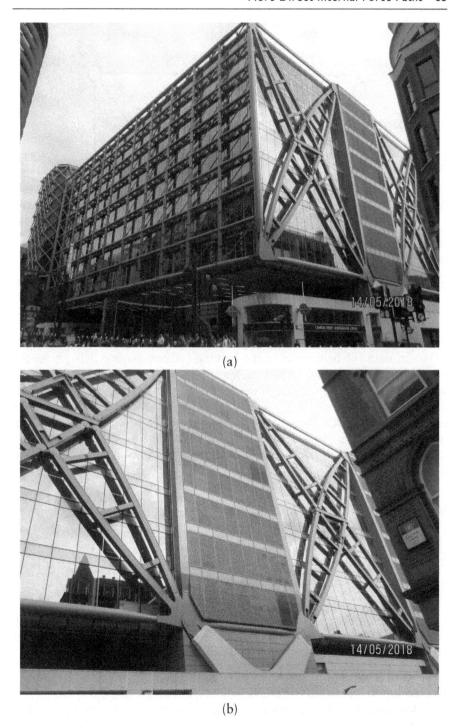

(a)

(b)

Figure 3.23 A building over the Cannon Street Underground Station, London. (a) The cantilevered upper eight-storey are supported by a huge bracing system. (b) The bracing members show the force paths to the foundation.

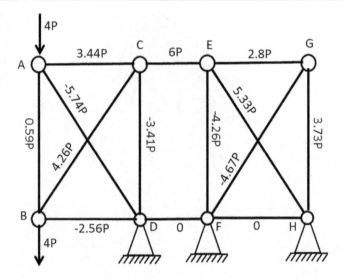

Figure 3.24 Paths and estimation of the internal forces in the truss system.

To estimate the behaviour of the bracing system, assume that the height to width ratio of the unit ABDC is 4/3. Approximate and replace the loading and self-weight of the cantilever part, ABDC, by two vertical loads, 4P, applied at nodes A and B (Figure 3.24). The values of internal forces are shown adjacent to members with positive signs for tension and negative sign for compression. In this case, the bracing members provide clear and desired internal force paths for transmitting the loads of the cantilever building to the supports.

References

3.1   Ji, T. and Ellis, B. R. Effective Bracing Systems for Temporary Grandstands, *The Structural Engineer*, 75(6), 95–100, 1997.

3.2   Ji, T. Concepts for Designing Stiffer Structures, *The Structural Engineer*, 81(21), 36–42, 2003.

3.3   Yu, X., Ji, T. and Zheng, T. Relationships Between Internal Forces, Bracing Patterns and Lateral Stiffness of a Simple Frame, *Engineering Structures*, 89, 147–161, 2015.

3.4   Roohi, R. *Analysis, Testing and Model Demonstration of Efficiency of Different Bracing Arrangements*, Investigative Project Report, UMIST, 1998.

3.5   Parkyn, N. *Super Structures: The World's Greatest Modern Structures*, Merrell, 2004.

3.6   Bennett, D. *Skyscrapers—Form & Function*, Simon & Schuster, 1995.

3.7   Billington, D. P. *The Tower and the Bridge*, Princeton University Press, Princeton, 1985.

3.8   Eley, D. and Annereau, N. The Structural Engineering of the Leadenhall Building, London, *The Structural Engineer*, 96(4), 10–20, 2018.

3.9   Saeed, M. *Parametric Study on the Diagrid Frame of the Leadenhall Building & Topology Optimisation of Bracing Systems*, MSc Dissertation, The University of Manchester, 2018.

3.10 Institution of Structural Engineers. *Temporary Demountable Structures: Guidance on Procurement, Design and Use*, Third Edition, Institution of Structural Engineers, London, 2007.

3.11 Ellis, B. R., Ji, T. and Littler, J. The Response of Grandstands to Dynamic Crowd Loads, *Structures and Buildings, the Proceedings of Civil Engineers*, 140(4), 355–365, 2000.

3.12 Anderson, J. Teaching Health and Safety at University, *Proceedings of the Institution of Civil Engineers, Journal of Civil Engineering*, 114(2), 98–99, 1996.

Chapter 4

Smaller Internal Forces

4.1 Routes to Implementation

There are several routes to create smaller internal forces which are apparent and intuitive. These provide a basis for developing particular implementation measures to allow the realisation of smaller internal forces in structures

1. Reducing Spans

As the deflection is proportional to the span to the power of four, reducing span whenever possible is the most effective way to achieve smaller deflections. For example, if the span of a beam is halved, the maximum deflection will be one sixteenth of that of the original beam. The maximum bending moment will also be reduced, to one quarter of that of the original beam.

2. Partially Self-Balancing Internal Forces

While it may not be possible to achieve complete self-balancing of internal forces, it may be possible to achieve partial self-balancing. The reduction of large bending moments can be realised by creating partially self-balanced systems in which a newly generated positive (negative) bending moment offsets part of an existing negative (positive) bending moment. To do this a designer needs to sense where the large bending moment would occur and its direction and then, more importantly, needs to develop an appropriate physical measure to introduce the new bending moment. Some measures which can be used to achieve partial self-balancing are:

1. Using pre-stressing or post-stressing techniques to produce a bending moment or deflection that is in the opposite direction to that induced by loads.
2. Adding structural elements into a structure which can constrain some deflections and/or create bending moments in the opposite direction to that induced by loads.
3. Redistributing internal forces which helps to reduce large bending moments.

3. Providing Elastic Supports

Providing rigid supports to reduce deflections may be difficult to be realised in practice due to functional, structural or aesthetic requirements. Using elastic supports may however be a feasible solution. There are two types of elastic supports: external elastic supports and internal elastic supports.

1. External elastic supports: When an elastic support is cut, a pair of action and reaction forces are revealed which have equal magnitudes but in opposite directions. If one force acts on the structure while the other does not act on the structure, this is an external elastic support. The ring shown in Figure 4.1a has a pair of horizontal springs to restrain its lateral deflection. The spring forces act on the ring and on solid supports that are not part of the ring. Therefore, these two springs act as external elastic supports to the ring. Typical structures making use of external elastic supports are the many cable stayed bridges in which the cables effectively act as elastic supports to the bridge decks and allow the bridge span longer distances and experience smaller bending moments. The other ends of the cables locate on the pylons that are supported by their foundations.
2. Internal elastic supports: If the action and reaction forces from an elastic support are all applied directly to the structure, it is regarded as an internal elastic support. The ring shown in Figure 4.1b has a tendon across its centre which acts as two springs to restrain the deflections in the lateral directions of the ring due to the applied load. The actions of the tendon on the ring are similar to those of the two external springs in Figure 4.1a but the forces in the tendon act on the two sides of the ring. The internal elastic supports provided by the tendon may also be considered as the realisation of partial self-balancing.

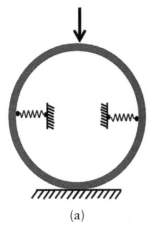
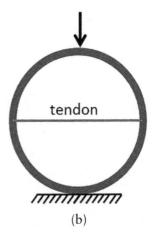

(a)   (b)

Figure 4.1 Rings with lateral elastic supports. (a) Springs acting as external elastic supports. (b) Tendon acting as internal elastic supports.

4.2 Hand Calculation Examples

4.2.1 A Simply Supported Beam with and without Overhangs

This example demonstrates and quantifies the effectiveness and efficiency of span reduction and self-balancing of internal forces through using overhangs.

Figure 4.2 shows three simple beams that have the same rigidity of EI. Beam 1 in Figure 4.2a is a normal, simply supported beam with a span of L and is subjected to a uniformly distributed load of q. The other two beams are evolved from Beam 1. When the two supports of Beam 1 are moved inward symmetrically with a distance of μL, it becomes Beam 2 (Figure 4.2b) that is called a beam with overhangs. When Beam 1 increases its overall length by αL at each of its two ends and a concentrated load of P is applied at each of its two free ends, it becomes Beam 3 as shown in Figure 4.2c, in which α and P can be variables to be determined for achieving a more efficient design. Determine the maximum bending moments

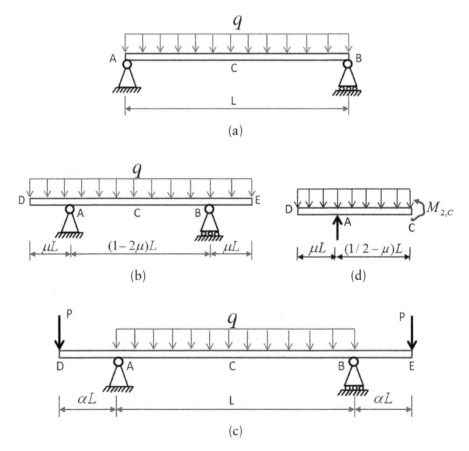

Figure 4.2 A simply supported beam and its two variations. (a) Beam 1: A simply supported beam. (b) Beam 2: A simply supported beam with overhangs and the same length as Beam 1. (c) A simply supported beam with overhangs for a total length of $(1 + 2\alpha) L$. (d) Free-body diagram used to determine the bending moment at the centre of Beam 2.

and the deflections at the centre C of the three beams and examine the efficiency of Beams 2 and 3 against Beam 1. Some basic formulae for calculating the bending moments and deflections of such beams can be found in related textbooks [4.1].

Solution:

Beam 1: A simply supported beam (Figure 4.2a).

The maximum bending moment and deflection at the centre of the simply supported beam are respectively:

$$M_{1,C} = \frac{1}{8}qL^2; \ \Delta_{1,C} = \frac{5qL^4}{384EI} \qquad (4.1a, 4.1b)$$

Beam 2: A simply supported beam with overhangs and with the same length as Beam 1 (Figure 4.2b).

The bending moments at supports A and B are:

$$M_{2,A} = M_{2,B} = -\frac{1}{2}q\mu^2L^2 \qquad (4.2a)$$

The bending moment at mid-span C of the beam can be determined using the free body diagram shown in Figure 4.2d as follows:

$$M_{2,C} = \frac{1}{2}qL\left(\frac{1}{2}-\mu\right)L - \frac{1}{2}q\left(\frac{L}{2}\right)^2 = \frac{1}{8}qL^2 - \frac{1}{2}q\mu L^2 \qquad (4.2b)$$

μ ($< \frac{1}{2}$) is a variable in equations 4.2a and 4.2b and can be adjusted to achieve smaller bending moments. Consider the particular case when the magnitudes of moments at location A and C, ($M_{2,A}$ and $M_{2,C}$) are the same. Equating the magnitudes in equations 4.2a and 4.2b gives:

$$\frac{1}{2}q\mu^2L^2 = \frac{1}{8}qL^2 - \frac{1}{2}q\mu L^2 \quad or \quad 4\mu^2 + 4\mu - 1 = 0 \qquad (4.3)$$

The valid solution of the quadratic equation in equation 4.3 is $\mu = 0.207$. Substituting $\mu = 0.207$ into the expressions for $M_{2,A}$ and $M_{2,C}$ leads to:

$$M_{2,C} = -M_{2,A} = \frac{1}{2}q(\mu L)^2 = \frac{1}{2}q(0.207)^2L^2 = 0.0214qL^2 \qquad (4.4a)$$

Alternatively, the bending moment at the centre of the beam can be determined as half of the maximum bending moment of the simply supported beam with the span of: $(1-2\mu)L = 0.586L$

$$M_{2,C} = -M_{2,B} = \frac{1}{2}\frac{1}{8}q[(1-2\mu)L]^2 = \frac{1}{16}q(0.586)^2L^2$$
$$= 0.0214qL^2 = 17.1\%M_{1,C} \qquad (4.4b)$$

The superposition method can be used to calculate the deflection at mid-span C. The loading in Figure 4.2(b) can be decomposed into two simple cases as shown in Figure 4.3a and Figure 4.3b. The deflection at C due to the distributed loads on the overhangs (Figure 4.3b) is the same as that due to two couples acting at the supports A and B (Figure 4.3c), which is $q(\mu L)^2 / 2$. This loading generates an upward deflection of the middle span, which can be calculated using a textbook equation [4.1] as:

$$
\begin{aligned}
\Delta_{2,C} &= \frac{5q[(1-2\mu)L]^4}{384EI} - \frac{M[(1-2\mu)L]^2}{8EI} \\
&= \frac{5q(0.586L)^4}{384EI} - \frac{q(0.207L)^2}{2}\frac{(0.586L)^2}{8EI} \\
&= 0.1179 \times \frac{5qL^4}{384EI} - 0.0706 \times \frac{5qL^4}{384EI} = 0.0473 \times \frac{5qL^4}{384EI} \\
&\qquad\qquad = 4.73\%\Delta_{1,C}
\end{aligned}
\tag{4.5}
$$

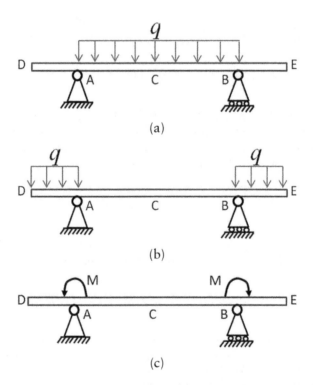

Figure 4.3 Using the superposition method for calculation. (a) Distributed loads applied between the two supports. (b) Distributed loads applied on the overhangs. (c) Equivalent loads to (b) for calculating the deflection at the mid-span.

The results show that the maximum bending moment for Beam 2 is about 17% of that for Beam 1 and that the maximum deflection for Beam 2 is less than 5% of that for Beam 1. Such significant reductions are due to the use of the first two physical measures in Section 4.1:

1. *The reduced span between the two supports*: Bending moment is proportional to the span squared and deflection is proportional to the span to the power four. Hence the shortened span effectively reduces the bending moment and deflection.
2. *The reduction of bending moment through part self-balancing*: The negative bending moments over the supports due to the use of the overhangs offset part of the positive bending moment due to the loads at the middle span. This can also be explained as a redistribution of bending moments. The reduced bending moment will also lead to smaller deflection.

The effects of the span reduction for Beam 2 can be seen when the distributed load is only applied on the middle span between the supports of Beam 2, for which the bending moment at the centre C is

$$M_{2,C} = \frac{1}{8}q[(1-2\mu)L]^2 = 0.586^2\frac{qL^2}{8} = 34.3\%M_{1,C} \tag{4.6}$$

Equation 4.6 indicates that the span reduction leads to a reduced maximum bending moment of 34.3% of $M_{1,C}$. If applied, the loading on the two overhangs (Figure 4.3b) further reduces the bending moment of 34.3% $M_{1,C}$ by a half to 17.1% (equation 4.4b). The first term in Equation 4.5 shows that the span reduction results in a reduction of 88.2% of $\Delta_{1,C}$ while the loading on the overhangs causes a further reduction of 7.06% of $\Delta_{1,C}$. Table 4.1 summarises the efficiency of the two measures on the reduction of both maximum bending moment and maximum deflection.

It can be seen from Table 4.1 that the measure of span reduction plays the dominant role in reducing the structural responses, in particular the deflection, and the self-balancing measure reduces the bending moment more significantly than deflection in this example.

Figure 4.4 shows the roof structure for the new entrance of the Fountains Abbey in North Yorkshire, UK. The roof is supported by a series of parallel curved beams that are in turn supported directly by individual columns

Table 4.1 Summary of the Efficiency on the Reduction of Bending Moment and Deflection

	Maximum bending moment	*Maximum deflection*
Span reduction	65.7%	88.2%
Self-balancing action	17.1%	7.06%
Total	82.8%	95.3%

Figure 4.4 Simply supported curved beams with overhangs.

at their lower ends and by a beam spanning between columns toward their upper ends. Examining the supports of the curved beams, this is an implementation of the simply supported beam with overhangs shown in Figure 4.2b. The structural behaviour of the curved beams in the vertical direction is the same as that of a straight beam with overhangs, but the curved roof surface and unequal heights of the supports are aesthetically pleasing to the eyes of visitors.

In engineering practice $\mu = 0.2$ is used rather than the exact solution of $\mu = 0.207$ for a simply supported overhanging beam for ease of design.

Beam 3: A simply supported beam with overhangs with a total length of $(1 + 2\alpha)L$ (Figure 4.2c).

Similar to the solution of Beam 2, the superposition method is used again to calculate the bending moments at A(B) and C and the deflections at C and D(E) of the beam.

The bending moments at B and C are respectively:

$$M_{3,B} = P\alpha L \tag{4.7}$$

$$M_{3,C} = \frac{1}{8}qL^2 - P\alpha L = \frac{1}{8}qL^2(1 - \frac{8P\alpha}{qL}) \tag{4.8}$$

where α is a design parameter and P can be either a given load or a design force parameter. The downward deflection at C due to the distributed load q alone is:

$$\Delta_{3,C1} = \frac{5qL^4}{384EI}$$

The upward deflection at C due to the two concentrated loads of P alone (calculated on the basis of a simply supported beam subjected to couples at A and B of $M = P \times \alpha L = \alpha PL$ (see Figure 4.3(c)) is:

$$\Delta_{3,C2} = \frac{ML^2}{8EI} = \frac{\alpha PL^3}{8EI}$$

Therefore, the downward deflection at C due to the full loading on the span of the beam is the sum of the two sub-loading case deflections:

$$\Delta_{3,C} = \Delta_{3,C1} - \Delta_{3,C2} = \frac{5qL^4}{384EI} - \frac{\alpha PL^3}{8EI} = \frac{(5qL - 48\alpha P)L^3}{384EI} \tag{4.9}$$

To determine the deflection at D or E, it is necessary to know the slopes of deformation induced by the two sub-loading cases. Due to the action of the distributed load alone, there is an upward deflection at D caused by the rotation of member DA, which can be calculated using an existing formula [4.1] as:

$$\Delta_{3,D1} = \theta_{3,A1} \times \alpha L = \frac{qL^3}{24EI} \times \alpha L = \frac{\alpha qL^4}{24EI}$$

The downward deflection at D due to the concentrated load is the sum of two deflections: the end deflection of a cantilever of length of αL due to P at its free end and the end deflection due to the rotation of the overhang DA:

$$\Delta_{3,D2} = \frac{PL^3}{3EI} + \theta_{3,A2}\alpha L = \frac{PL^3}{3EI} + \frac{(P\alpha L)L}{2EI}\alpha L = \frac{PL^3(2 + 3\alpha^2)}{6EI}$$

The total downward deflection at D is the sum of the deflections due to the two sub-loading cases, *i.e.*:

$$\Delta_{3,D} = \Delta_{3,D2} - \Delta_{3,D1} = \frac{PL^3(2 + 3\alpha^2)}{6EI} - \frac{\alpha qL^4}{24EI} = \frac{[4P(2 + 3\alpha^2) - \alpha qL]L^3}{24EI} \tag{4.10}$$

Equations 4.7 to 4.10 contain two variables, P and α, and these variables can be used to actively adjust the bending moment and deflection of the beam with overhangs for the loading condition shown in Figure 4.2c.

Figure 4.5 shows a steel-framed two-storey car park building, which embodies the study of Beam 3. The vertical loads from floors are transmitted to the cellular

Figure 4.5 Overhangs and tendon forces are used to reduce bending moments and deflections of cellular beams (Courtesy of Mr John Calverley, UK).

beams and then from the cellular beams to the supporting columns. Overhangs are purposely designed in the structure to reduce the bending moments and deflections of the cellular beams. Examining the first overhang, two steel cables link the free end of the overhang and a concrete support. A downward force at the free end of the overhang is provided by tensions induced in the cables. This force, similar to P in Figure 4.2c, will generate a negative bending moment in the beam over the column support which will partly offset the positive moments in the beam induced by the floor loading. The length of overhangs and the force in the steel cables could be the design parameters actively selected to reduce the bending moments and deflections of the cellular beam.

The overhang is subjected to a concentrated force at its free end and therefore the bending moment varies linearly along the overhang from zero at its free end to a maximum at the column support. Reflecting the shape of the bending moment diagram, the overhang is tapered toward the column. This makes the overhang appear lighter and more elegant than would be the case if a constant cross-section was used throughout its length. A prop is provided between the concrete support and the column end of the overhang which stiffens the overhang to prevent its rotational deformation due to the action of the cables, contributes additional lateral resistance to the structure and provides anchoring positions for the cables.

4.2.2 Y Shaped Columns with and without a Horizontal Tendon

This example demonstrates and quantifies the effectiveness and efficiency of self-balancing through the use of a bar member in a Y shaped column.

Figure 4.5 shows two Y shaped columns, one without and one with a horizontal bar linking the two top ends of the column, which are subjected to the same pair of symmetric vertical loads. The dimensions of the Y shaped column can be described by three parameters: the column height, h, the span, a, and height, b, of the two symmetric inclined members. The length of the inclined member is $s = \sqrt{a^2 + b^2}$. Assume that the Y shaped columns have a uniform cross-section with a rigidity of EI and that the horizontal bar has a sectional area of A and elastic modulus of E_b. Conduct the following analyses:

1. Determine the bending moments in the two Y shaped columns.
2. Determine the vertical deflection at point A and the relative horizontal deflections between points A and B of the two inclined members.
3. Examine the effect of the horizontal bar on the reduction of the bending moment and lateral and vertical deflections of the Y shaped columns.

Solution:

Column 1 (Figure 4.6a):

The Y shaped column is a statically determinate structure and its bending moment diagram can be drawn easily as shown in Figure 4.7a. There is no bending moment in the vertical column as the moments induced by the pair of symmetric vertical loading are self-balancing at the connection point C.

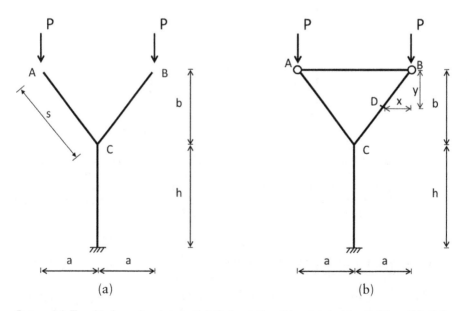

Figure 4.6 Two Y shaped columns. (a) Column 1, without a horizontal bar. (b) Column 2, with a horizontal bar.

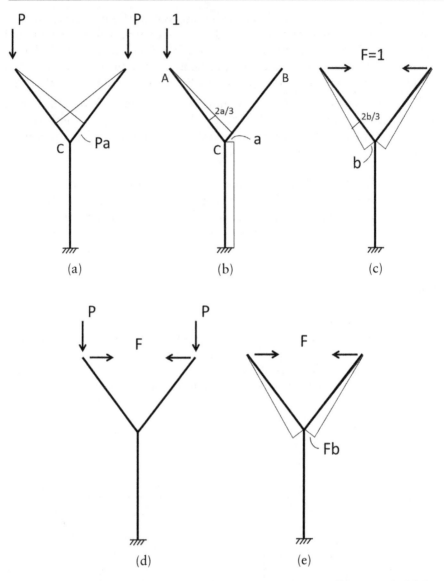

Figure 4.7 Bending moment diagrams of the Y column due to different loads. (a) Due to a pair of vertical loads. (b) Due to a unit downward load at point A. (c) Due to a pair of unit horizontal inward forces. (d) Releasing the bar force in the tied Y shaped column to make it a statically determinate structure. (e) Due to a pair of horizontal inward forces F.

The maximum bending moment occurs at point C and has a magnitude of $M_{1,\max} = Pa$.

The vertical and horizontal deflections can be calculated using the unit load method and the moment-area method in conjunction with Figures 4.7a, 4.7b and 4.7c, which show the bending moment diagrams for the Y shaped column due to the pair of vertical loads P, a unit downward vertical load acting at point A and due to a pair of inward horizontal unit forces at points A and B.

The vertical downward deflection of point A induced by the pair of loads P is (Figure 4.7a and 4.7b):

$$\Delta_{1,V} = \frac{1}{EI}(\frac{Pas}{2})(\frac{2a}{3}) = \frac{Pa^2s}{3EI} \tag{4.11}$$

The relative horizontal deflection between points A and B is:

$$\Delta_{1,H} = \frac{1}{EI}(\frac{Pas}{2})(-\frac{2b}{3})\cdot 2 = -\frac{2Pabs}{3EI} \tag{4.12}$$

In equations 4.11 and 4.12, the first subscript specifies the Y shaped column (Figure 4.6) and the second indicates the direction of the deflection. The negative sign in equation 4.12 means that the relative horizontal deflection between points A and B is opposite to the direction of the unit horizontal forces shown in Figure 4.7c, *i.e.* points A and B deform outwards.

Column 2 (Figure 4.6b):

The Y shaped column with a horizontal bar is a statically indeterminate structure as the internal force in the bar is unknown. The moment-area method can be used to determine the internal force F in the bar. When the bar is replaced by a pair of forces F as shown in Figure 4.7d, the Y shaped column becomes statically determinate. Figure 4.7e shows the bending moment diagram due to the pair of horizontal forces, F. The force, F, can be determined using the deflection compatibility condition between points A and B of the tied Y shaped column as follows.

The horizontal deflection due to the pair of vertical loads P (Figures 4.7a and 4.7c) is:

$$\Delta_{H,P} = \frac{1}{EI}(\frac{Pas}{2})(-\frac{2b}{3})\cdot 2 = -\frac{2Pabs}{3EI} \tag{4.13}$$

The horizontal deflection due to a pair of horizontal forces F (Figures 4.7e and 4.7c) is:

$$\Delta_{H,F} = \frac{1}{EI}(\frac{Fbs}{2})(\frac{2b}{3})\cdot 2 = \frac{2Fb^2s}{3EI} \tag{4.14}$$

The elongation of the horizontal bar is:

$$\Delta_b = \frac{F\cdot 2a}{E_b A} \tag{4.15}$$

Compatibility of the deflections in equations 4.13, 4.14 and 4.15 requires:

$$\Delta_{H,P} + \Delta_{H,F} + \Delta_b = 0 \tag{4.16}$$

Substituting equations 4.13 to 4.15 into equation 4.16 gives:

$$-\frac{2Pabs}{3EI} + \frac{2Fb^2 s}{3EI} + \frac{2Fa}{E_b A} = 0 \qquad (4.17)$$

The signs in the three deflections in equations 4.13 to 4.15 may be a little confusing, but they can be judged from an understanding of the physical nature of the deflections of the Y shaped column without the horizontal bar. The action of the pair of vertical loads shown in Figure 4.7a alone causes points A and B to deflect outwards, while the deflection due to the bar forces, F, alone (Figure 4.7e) would be inwards and smaller than that due to P. The difference between two deflections is the elongation of the bar, i.e.:

$$\left| \Delta_{H,P} \right| - \left| \Delta_{H,F} \right| = \left| \Delta_b \right|$$

which is effectively what equation 4.17 states.

There is only one unknown, F, in equation 4.17 and solving for F gives:

$$F = \frac{Pa}{b} \frac{1}{(1 + \dfrac{3EIa}{b^2 s E_b A})} = \frac{Pa}{b} k \qquad (4.18)$$

$$k = \frac{1}{1 + \dfrac{3EIa}{b^2 s E_b A}} \qquad (4.19)$$

where k is smaller than 1 and is a non-dimensional coefficient that is related to the geometry and cross-sectional properties of the inclined member of the Y shaped column and the properties of the horizontal bar. If the rigidity of the bar, $E_b A$, becomes infinite, then $k = 1$ and the tension force in the bar becomes $F = Pa / b$. For this scenario, there are no bending moments in any of the members making up the tied Y shaped column when subjected to the symmetric vertical loads, i.e. the bending moments induced by P are balanced by the bending moments induced by the horizontal force F in the rigid bar. This can be demonstrated by calculating the bending moment at any point D of the right aim of the tied Y column (Figures 4.6b and 4.7d) as follows:

$$M_D = P \cdot x - F \cdot y = Px - \frac{Pa}{b} \times \frac{b}{a} x = 0 \qquad (4.20)$$

While it is impractical that the bar rigidity $E_b A$ could be infinite, this corresponds to the equivalent situation that the lateral deflections of the two top ends of the Y shaped column are constrained by roller supports in the horizontal direction.

Once the internal force of the bar, F, has been determined, the Y shaped column with a horizontal bar becomes a statically determinate structure as shown in Figure 4.7d and the bending moment and deflections at the key positions can be easily calculated.

The maximum bending moment in an inclined member is:

$$M_{2,\max} = Pa - Fb = Pa - Pak = Pa(1-k) = M_{1,\max}(1-k) \tag{4.21}$$

The vertical downward deflection at A due to P and F is:

$$\begin{aligned}
\Delta_{2,V} &= \frac{1}{EI}(\frac{Pas}{2})(\frac{2a}{3}) - \frac{1}{EI}(\frac{Fbs}{2})(-\frac{2a}{3}) = \frac{Pa^2s}{3EI} - \frac{Fabs}{3EI} \\
&= \frac{Pa^2s}{3EI} - \frac{Pa^2ks}{3EI} = \frac{Pa^2s}{3EI}(1-k) = \Delta_{1,V}(1-k)
\end{aligned} \tag{4.22}$$

The horizontal outward deflection between A and B due to the action of P and F is:

$$\begin{aligned}
\Delta_{2,H} &= \frac{1}{EI}(\frac{Pas}{2})(\frac{2b}{3}) \times 2 - \frac{1}{EI}(\frac{Fbs}{2})(-\frac{2b}{3}) \times 2 = \frac{2Pabs}{3EI} - \frac{2Fb^2s}{3EI} \\
&= \frac{2Pabs}{3EI} - \frac{2Pabks}{3EI} = \frac{2Pabs}{3EI}(1-k) = \Delta_{1,H}(1-k)
\end{aligned} \tag{4.23}$$

Equations 4.18 and 4.19 show that F is smaller than Pa/b as the coefficient k is less than 1. When the cross-sectional properties, I and A, are measured in meters, the value for *area A* would be much larger than that of I; a, b and s are geometric dimensions of the cantilever arms and $s > a$. Therefore, the ratio $3EIa/(b^2sE_bA)$ is likely to be much smaller than 1.0 for most practical cases. Consequently, the coefficient k in equation 4.19 would not be much smaller than 1. Equations 4.21 to 4.23 show that the maximum bending moment at C and the vertical and horizontal deflections at node A of the tied Y shaped column are $(1-k)$ times of that of the same Y shaped column without the horizontal bar.

In order to gain a feel for the effect of the horizontal bar on the maximum bending moment and deflections of a Y shaped column, a particular case with the following data is examined.

The inclined member of the Y shaped column has dimensions of $a = 2.0m$, $b = 1.5m$ and $s = 2.5m$ and uses an I section steel beam, UB254 x 102 x 25, with a second moment of area of $I = 3415\,cm^4 = 3.415 \times 10^{-5}\,m^4$. The steel bar has a radius of 1.0 cm, i.e. a cross-sectional area of $A = 3.14\,cm^2 = 3.14 \times 10^{-4}\,m^2$. The elastic modulus for both inclined members and the bar are the same with $E = E_b = 200 \times 10^9\,N/m^2$. Vertical loads of 100kN act on points A and B (Figure 4.6b).

Coefficient k is thus:

$$k = \frac{1}{(1+\dfrac{3Ia}{b^2 sA})} = \frac{1}{(1+\dfrac{3\times 3.415\times 10^{-5}\times 2}{1.5^2 \times 2.5 \times 3.14\times 10^{-4}})} = \frac{1}{1+0.116} = 0.896$$

The horizontal force in the steel bar can be determined using equation 4.18 as:

$$F = \frac{Pa}{b}k = \frac{100,000\times 2}{1.5}\times 0.896 = 119,467N$$

The bending moment at C is:

$$M_{2,\max} = Pa(1-k) = 100,000\times 2.0 \times (1-0.896)$$
$$= 200,000\times 0.104 = 20,800Nm$$

The vertical deflection at A, from equation 4.22 is:

$$\Delta_{2,V} = \frac{Pa^2 s}{3EI}(1-k) = \frac{100,000\times 2^2 \times 2.5}{3\times 200\times 10^9 \times 3415\times 10^{-8}}(1-0.896)$$
$$= 0.0488\times 0.104 = 0.0051m$$

The relative deflection between A and B based equation 4.23 is:

$$\Delta_{2,H} = \frac{2Pabs}{3EI}(1-k) = \frac{2\times 100,000\times 2\times 1.5\times 2.5}{3\times 200\times 10^9 \times 3415\times 10^{-8}}(1-0.896)$$
$$= 0.0732\times 0.104 = 0.0076m$$

It can be observed that the use of the horizontal bar effectively controls both horizontal and vertical deflections of the two inclined members of the Y shaped column, which leads to much smaller internal forces and deflections. For this particular case, the reductions are significant, up to about 90% of similar values for the normal Y shaped column. Therefore, it can be said that the horizontal bar creates internal elastic supports to the tops of the Y shaped column which leads to smaller deflections and internal forces. Alternatively, it can be explained as the bending moments induced by the horizontal bar partly balance those induced by the vertical loads, which results in much smaller internal forces and consequently smaller deflections. These explanations indicate that the physical measure of using a horizontal bar to tie the two top ends of the Y shaped column can be generated from different ways of thinking.

Figure 4.8 shows two practical examples in which tied Y shaped columns have been used in past and present times. The tied Y shaped columns in Figure 4.8a are in the railway station in Knaresbough, North Yorkshire, UK. The station was built in 1890, and the Y shaped columns were made of cast iron.

(a)

Horizontal bar

(b)

Figure 4.8 The use of Y shaped columns with a horizontal bar. (a) At a train station. (b) At an airport terminal.

It can be seen in Figure 4.8a that the tie member is in fact the lower chord of a roof truss. This has a thicker section than that of the two arms which effectively prevents the ends of the curved arms from deforming horizontally and vertically. The use of curved arms instead of the conventional straight arms is more aesthetically pleasing.

Figure 4.8b shows a straightforward implementation of the Y shaped column with a horizontal bar. This steel tied Y shaped column is in Terminal 5 at the Heathrow Airport, London, which was opened in 2008. It can be noted that the horizontal bar has a small cross-section in comparison with that of the two inclined members. The dimensions and properties of the Y shaped column were estimated and used in the hand calculations. The use of the horizontal bar was seen to reduce about 90% of the deflections and bending moments in a similar Y shaped column without a horizontal bar.

The main differences between the tied Y shaped columns shown in Figure 4.8 are the materials used and the technology involved. In spite of the differences in locations, construction times and materials, the structural concept embedded in the two designs is essentially the same indicating that the implementation of structural concepts is not dependent on time or material.

4.3 Practical Examples

4.3.1 Structures with Overhangs

4.3.1.1 HSBC Hong Kong Headquarters, China

The beam with overhangs discussed in Section 4.2.1 is simple and efficient, and the embedded physical measures of reducing span and partly self-balancing bending moments can be applied to more advanced structures. Figure 4.9(a) shows the tower of the HSBC Hong Kong headquarters, which has 47 storeys, stands 179m above ground and was built between 1979 and 1986 [4.2, 4.3]. Figure 4.9b is a model of the building.

The main structure of the building is exposed allowing for direct appreciation. The building structure is supported by eight masts, arranged in two rows of four (two masts can be seen in Figure 4.9a). Each mast consists of four tubular steel columns which are rigidly connected by rectangular beams and supported on foundations driven into bedrock over 30m below ground level. Bracing members are provided between the masts which effectively increase the lateral stiffness of the building structure. The masts support five discrete, double, two-storey height, steel trusses which span 33.5m between the masts and cantilever 10.7m beyond them. This mast and truss system carries all the structural loads and creates a spectacular column-free area at ground floor level. Each truss supports several lower floors on hangers at the centre and at the two ends of the truss. Figure 4.10 shows clearly the top ends of the central and side hangers, indicating that the truss supports the floors below. One of the discrete truss systems is now chosen for a closer examination of the load (internal force) paths.

(a)

Figure 4.9 The tower of HSBC Hong Kong headquarters. (a) A front view. (b) The model.

Due to the symmetry in elevation and for easy understanding, Figure 4.11 shows half of the elevation of the structure consisting of two columns with link beams and the five storeys of floor beams supported from a truss by two hangers (CG and FH). Two roller supports are provided to reflect the symmetry and to prevent lateral deflections. The main floor beams are pin-connected to the hangers and to the columns. The vertical loads acting on the main beams at floor levels are transmitted to the columns and hangers which generate compression forces in the columns and tension forces in the hangers. The tension forces in the hangers are then transmitted to nodes C

(b)

Figure 4.9 (Continued)

and F, and then to members BF and AC in tension and CD and EF in compression. It can be appreciated that the horizontal component of the force acting on node B from member BF is partly self-balanced by the force acting on node A from member AC. Similarly, the horizontal force acting on node E from member EF is partly balanced by the force acting on node D from member CD. The effect of the self-balancing effectively reduces the lateral forces on the masts which consequently reduces the bending moments in the masts.

Figure 4.11 indicates that the floor beams can be analysed as individual, simply supported beams. Half of the loads on the beams are transmitted to the supporting hanger and the other half to the mast. For a better understanding of the structural behaviour and the effect of self-balancing due to

Figure 4.10 The truss supports the floors underneath rather than those above.

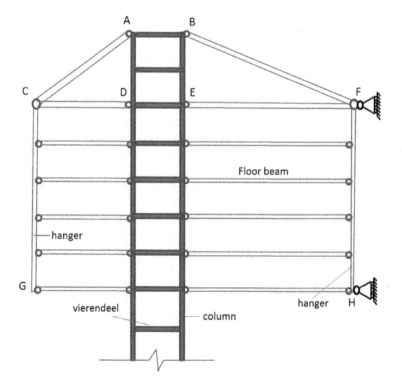

Figure 4.11 Illustration of a half of the elevation of the structure supported by one truss [4.3].

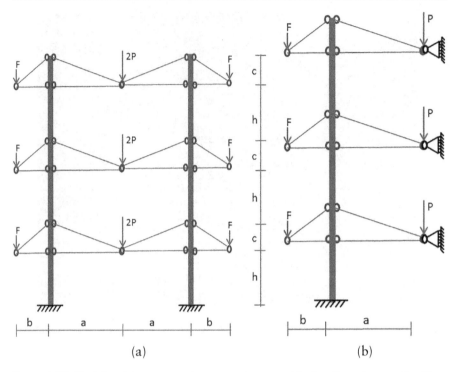

Figure 4.12 Models of the mast and truss system considering three trusses for illustration. (a) The full model in which the action of hangers is represented by point loads. (b) Simplified model based on symmetry.

the vertical loads, the structure can be further simplified to capture its physical essence for a hand calculation. Three levels of trusses and masts are considered as shown in Figure 4.12(a) with vertical loading applied through the hangers, $2P$ from the central hangers and F from the side hangers. When a symmetric structure is subjected to symmetric loads, the responses of the structure will be symmetric and hence in this case a half of the structure shown in Figure 4.12(b) can be considered, in which the central points of the trusses are constrained to prevent from any horizontal movements, reflecting the symmetry of deflection.

When the horizontal forces generated from the horizontal constraints in Figure 4.12b are not considered for estimation, the analysis of the model in Figure 4.12b becomes straightforward as the structure is statically determinate. Examining the lateral forces acting on the mast from the truss members (Figure 4.13a), there are six pairs of forces and that each pair of forces acts at the same level but in opposite directions. After partial self-balancing of the forces, six parallel forces are left and form three equal pairs of forces at different levels (Figure 4.12b). The corresponding shear

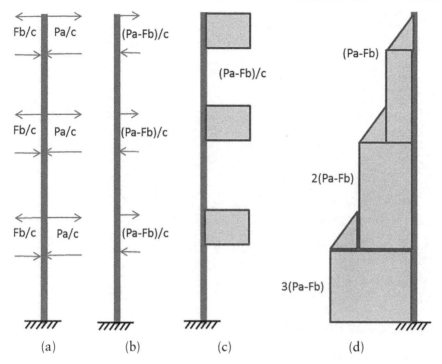

Fb/c   Pa/c
(Pa-Fb)/c

(Pa-Fb)/c

Fb/c   Pa/c
(Pa-Fb)/c

(Pa-Fb)

2(Pa-Fb)

Fb/c   Pa/c
(Pa-Fb)/c

3(Pa-Fb)

(a)   (b)   (c)   (d)

Figure 4.13 Loading and internal forces on the mast. (a) Loading from truss members. (b) Forces after self-balancing. (c) Shear force diagram. (d) Bending moment diagram.

force diagram and bending moment diagram for the mast are shown in Figures 4.13c and 4.13d.

The magnitudes of the horizontal forces acting on the column are $(Pa - Fb)/c$ and the self-balancing of forces is reflected in the term $Pa - Fb$, in which force F can be designed to achieve a more efficient structure. The qualitative structural model of the HSBC Hong Kong headquarters shown in Figure 4.12a is an extension of a simply supported beam with overhangs (Figures 4.2b and 4.2c) discussed in Section 4.2.1. Using overhangs reduces spans and creates partial self-balancing of internal forces.

4.3.1.2 Roof of the Harbin Airport Lounge, China

Figure 4.14 shows a roof structure with overhangs used in the lounge of the Harbin Airport terminal in China. The roof is supported by a series of trusses that are in turn supported by circular columns. The columns are positioned with a distance from the ends of the trusses and make the trusses to work like beams with overhangs discussed in Section 4.2.1.

Figure 4.14 A roof structure with overhangs in an airport terminal, Harbin, China.

4.3.2 Tree-Like Structures

4.3.2.1 Trees and Tree-Like Structures

The Y shaped column shown in Figure 4.5a is perhaps the simplest tree-like structure. Tree-like structures, also called branching structures, are structural forms developed from Y shaped columns with the addition of further branches. Trees, exposed to sun, rain, wind and other environment conditions, are so natural, logical and beautiful. Observing an oak tree in a winter (Figure 4.15) it can be seen that: 1) there is a structural hierarchy with the trunk thickest at the root of the tree and branches become thinner further away from the trunk. 2) The tree works effectively as a cantilever, *i.e.* the trunk is a vertical cantilever and the many individual branches act as smaller cantilevers. As a cantilever, it transmits the loads acting on it through bending, *i.e.* the bending moment becomes the largest at the end of the cantilever and gets smaller toward its tip. The thicknesses of the trunk and the branches of the tree basically reflect the relative magnitudes of the bending moments that are experienced.

The inherent beauty and natural forms of trees have been used and improved in architectural and structural designs in at least two ways:

1. The ends of branches have been used to support roofs or upper structures. Due to the supports provided by the branches the roofs or upper structures are able to span longer.

Figure 4.15 An oak tree in winter showing the structural hierarchy

2. The ends of branches have been linked by structural members in the roofs or upper structures which they support. Therefore, the branches no longer act as cantilevers, so with the end deflections of the branches constrained by the linking members they carry mainly axial forces rather than bending moments, which improves the efficiency and behaviour of the branch members. This feature has been illustrated using a Y shaped column with and without a tie in Section 4.2.2.

Section 4.2.2 demonstrates that the Y shaped column with a horizontal bar at its two top ends has much smaller bending moments than a Y shaped column without the bar under symmetric vertical loads, as the outward deflections of the two top ends of the column are constrained by the bar. The Y shaped column with a horizontal bar retains the tree like shape but is far more efficient, leading to the use of smaller cross-sections for the branches.

Consider the behaviour of a structure formed by a series of linked Y shaped columns as shown in Figure 4.16a, in which rotational and horizontal constraints are provided at the two top end nodes of the frame. This is a highly statically indeterminate structure, and simple hand calculations cannot be used directly. However, the structural form and the loads are both symmetric, and the property of symmetry can be utilised to simplify the structure and its analysis. The structure shown in Figure 4.16a can be represented as an equivalent half structure as shown in Figure 4.16b which is still symmetric. It can

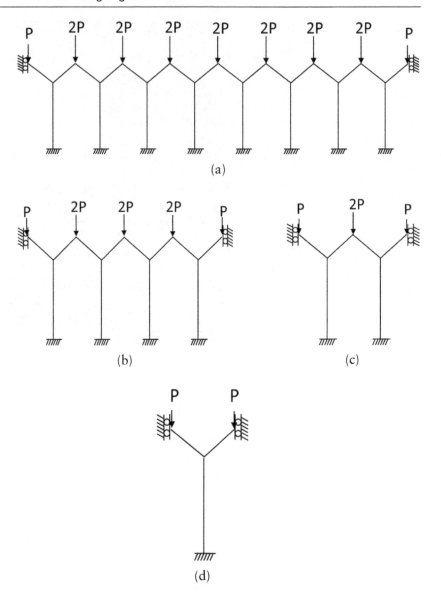

Figure 4.16 From a Y shaped frame to an equivalent single Y shaped column based on symmetry. (a) A series of Y shaped columns forming a frame structure. (b) Equivalent frame to (a) based on symmetry. (c) Equivalent frame to (b). (d) Equivalent frame to (c).

be similarly further simplified to a quarter of the original structure as shown in Figure 4.16c, then to the single Y shaped column shown in Figure 4.16d. Due to symmetry, this single Y shaped column with the rotational and horizontal constraints has two redundant forces, the lateral force and the bending moment, at the top end supports. The support forces can be determined by hand calculation, similar to the solution for the tied Y shaped column in

Figure 4.6b. The calculation shows that the pair of horizontal redundant forces equal Pa/b acting inward to each other while the bending moments at the supports are zero. Therefore, this constrained Y shaped column (Figure 4.16d) is equivalent to the tied Y shaped column (Figure 4.6b) when the rigidity of the horizontal bar has an infinite value. Equation 4.20 shows that the bending moment at any point in the inclined members become zero. *i.e.* the bending moment induced by vertical force P is completely offset by that due to the horizontal support force Pa/b. Because the structures shown in Figures 4.16a and 4.16d are the same, *i.e.* one can be generated from the other using symmetry, the members of the continuous Y frame structure (Figure 4.16a) do not experience any bending moments due to the given loads. This zero-moment scenario is created by constraining the lateral deflections of the top nodes of the Y shaped column. It is noted that deformations due to axial forces are negligible.

It is well known that a parabolic arch subjected to a uniformly distributed vertical load experiences no bending moment. This continuous Y shaped frame structure to the given loads (Figure 4.16a) is another example of a structure in which all members have no bending moment. The former case is a single structural member while the latter is a frame structure consisting of several members.

4.3.2.2 Palazzetto dello Sport, Roma

The use of Y shaped columns can be seen in the structure of the Palazzetto dello Sport (Small Sport Palace), shown in Figure 4.17, which was built in Rome in 1957 and engineered by Pier Luige Nervi. Figure 4.17 shows the form of the structure in which the shell roof is supported by a series of circularly arranged inclined Y shaped columns. Vertical columns are used to provide propping

Figure 4.17 Outlook of the Palazzetto dello Sport (Courtesy of Mr. Nicolas Janberg, structurae.net, Germany).

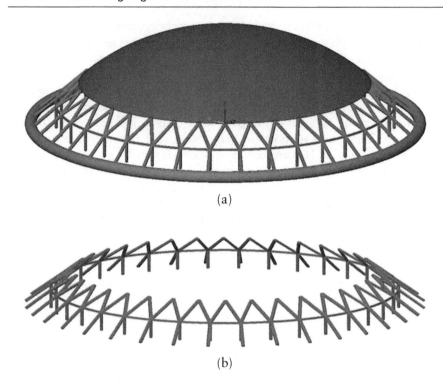

(a)

(b)

Figure 4.18 FE model of the Palazzetto dello Sport [4.4]. (a) FE model showing the three main parts of the structure, the roof, the series of Y shaped columns and the tension ring. (b) The isolated circularly enclosed Y frame structure.

supports to all the inclined Y shaped columns. Figure 4.18 shows a Finite Element (FE) model of the structure and the isolated circularly arranged Y shaped columns. As the exact data for the cross-sections of members of the structure are not available, the FE model provides an illustration for qualitative understanding of relative performance when several parameters of the structure are altered.

Due to the axisymmetry of the closed form of the series of Y frames to the vertical line through the centre of the roof and their connections to the roof shell, there is little lateral deflections occurring at the top ends of the Y shaped columns when subjected to uniformly distributed vertical loads, such as the self-weight of the roof structure. Following the qualitative analysis of the series of Y shaped columns in Figure 4.16 and the quantitative analysis of the single Y shaped column in Figure 4.6b, it can be inferred that there should be little bending moments in the planes of the Y shaped columns.

4.3.2.3 Hessenring Footbridge, Germany

The Hessenring Footbridge, designed by Schlaich Bergermann Partner, has a span of 46m and a width of 6.9m, and is located in Bad Homburg, Germany, Figure 4.19. The slender bridge deck is suspended by 16 cables that transfer

Figure 4.19 Hessenring Footbridge, Bad Homberg, Germany (Courtesy of Mr. Per Waahlin, Sweden).

the loads on the bridge deck to a tree-like mast located at the centre of the bridge. This mast is not only a loadbearing element of the bridge but a delicate, three-dimensional sculpture as well. The bifurcation of four arms from the truck, the central column, is an evolution of the Y column in two perpendicular directions. The four horizontal members linking the four ends of the arms confine the outward deflections of the arms due to the actions of the 16 cables, by which the four arms are mainly subjected to axial forces rather than bending moments. This deduction can come not only from the analysis of the Y shaped column with a tendon in section 4.2.2 but also from the observation that the arms have a similar cross-section along their lengths. If the four horizontal members were removed, the four arms would act like cantilevers with concentrated cable forces at their free ends. This would generate the bending moments in a triangular shape along the lengths of the arms with zero at their top ends and the maximum at their bottom ends.

4.3.2.4 Further Examples

Y shaped columns can be arranged in three dimensions to form tree-like structures. Figure 4.20 shows the structure of the Gare do Oriente Station in Lisbon, designed by Santiago Calatrava, where the branches of the trees are curved rather than straight. This variation would not affect the bending moments in the members of the columns under uniformly distributed vertical loads. In

Figure 4.20 A tree-like structure as an evolution of a series of Y shaped columns.

addition, the folded roof and many thin elements between branches constrain the relative deformation between the branches and between the columns, leading to only small bending moments. Therefore, no thick vertical columns and branch members are required in this structure.

Tree-like structures have been used creatively, and many variations have been produced to achieve aesthetic beauty and structural efficiency. As the tree-like structures have fewer columns but many more branches, they are able to provide good supports to roofs and are particularly suitable to be used in open spacious areas. Therefore they are often seen in shopping malls, exhibition centres and airport terminals. Figure 4.21a shows the Pu Dong Airport Terminal, Shanghai, in which the roof is supported by a series of Y shaped columns. The first level of branches of the Y shaped column is further divided into a second level of branches perpendicular to the ones in the first level creating four point supports for the roof. The connections between the roof and the tops of the Y shaped columns restrain the horizontal and vertical deformations of the four branches due to vertical loads. Thus, the members of the Y shaped columns are subjected mainly to compressive forces rather than bending moments resulting in lighter sections. The appearance of the thick vertical columns in Figure 4.21a is the use of the additional non-structural materials for architectural reasons and for the protection to the passengers.

Figure 4.21b shows the huge Y shaped columns used in the Bihai Cultural Centre in Tianjin, China, to support the roof over a large open area. It can be noted that eight branch members are developed from the central column to support the roof structure and that the top ends of the branches are connected to rigid roof members. The branch members are thus constrained to deform in the horizontal and vertical directions leading to only low bending moments in the branch members for vertical loads.

In the Madrid Barajas Airport terminal there are intensive uses of V shaped struts and Y shaped columns to support its roof structure, which allow the roof spanning over large areas without intermediate supports. Figure 4.22 shows the internal and external inclined Y shaped columns. The common feature of

(a)

(b)

Figure 4.21 Y shaped columns used for large public buildings. (a) Y shaped columns with two levels of branch used in Pu Dong Airport Terminal. Shanghai, China. (b) Large Y shaped columns used in the Bihai Cultural Centre, Tianjin, China (Courtesy of Mr. Peixuan Xie, UK)

(a)

(b)

Figure 4.22 Inclined Y shaped columns in the Madrid Barajas Airport terminal, Spain. (a) Internal use (Courtesy of Professor Zhaohui Chen, Chongqin University, China). (b) External use (Courtesy of Professor Guy Warzée—Université Libre de Bruxelles, Belgium).

these Y columns is that horizontal members are placed between the top ends of the Y columns. These horizontal members confine the lateral deflection of the ends and make the arms to subject mainly the axial forces.

4.3.3 Self-Balancing

4.3.3.1 Madrid Racecourse, Spain

Figure 4.23 shows the stand at the Madrid Racecourse (the Zarzuela Hippodrome), and a cross-section drawing and a physical model are shown in Figure 4.24. Figure 4.24 shows that the stand consists an upper roof or canopy, a seating area on the left and a betting hall on the right, which is covered by a lower roof. The upper roof is supported by central columns with pinned connections and rods, CD, between the upper and lower roofs. The left-hand side of the lower roof is rigidly connected to the central columns and is suspended at mid span by the rods, CD [4.5].

There are several merits of the structural design of the stand, but the partial self-balancing system embedded in the structure is of interest here. It is noted that rod CD (Figure 4.24a) is placed between the upper roof for the stand and the lower roof for the betting hall. The upper roof is supported by the vertical upward forces from the central columns and the downward forces from tensions in the rods. The weight of the lower roof is largely carried by the rods due to their locations at about the mid span of the lower roof. As the rods connect

Figure 4.23 A front view of the Madrid Racecourse Stand, Spain.

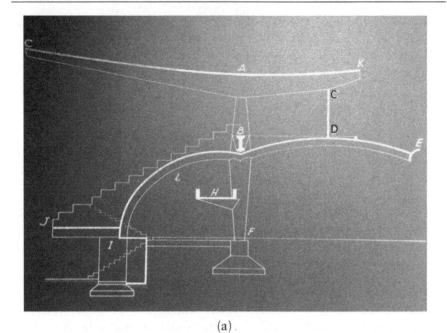

(a)

(b)

Figure 4.24 The design of the Madrid Racecourse Stand. (a) Cross-section drawing. (b) Physical model.

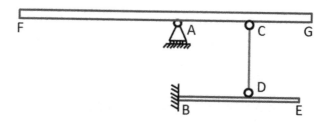

Figure 4.25 A simplified model of the two roofs and the rod linking them.

the upper and lower roofs, the two roofs and the rods form a partially self-balanced system. The interpretation and the physical essence of the system can be illustrated and investigated using the diagram in Figure 4.25 in which the relationships between the upper and lower roofs and the rod CD are presented. The upper roof requires downward forces to achieve its equilibrium while the lower roof needs upward forces to stiffen it and reduce its internal forces and deflections. The placement of rods, CD, serves the two purposes and makes that the upper roof FG and the lower roof BE are mutually supported.

4.3.3.2 Salford Quays Lift Bridge, UK

Using self-balancing to solve challenging engineering problems often achieves efficient designs. Figure 4.26 shows a lifting-up position of the Salford Quays lift bridge, also known as the Salford Quays Millennium Footbridge or the Lowry Bridge, that spans the Manchester Ship Canal between Salford and Trafford in Greater Manchester, England. The 91.2m long vertical lift bridge has a lift of 18m, allowing large watercraft to pass beneath.

The bridge consists of a pair of inward inclined arches that are connected at the crown, a bridge deck and a series of cables that are uniformly spaced along the length of the deck and linking the arch and the deck. Most of the self-weight of the deck and the traffic loads applied on the deck are transmitted to the cables then to the arches. An arch is effective to transmit external loads through mainly compression to its supports. However, it requires strong, substantial supports to balance the large horizontal forces at the ends of the arch. For this bridge (Figure 4.26), the ends of the arches and the deck are rigidly integrated and thus the deck, which is a bending member, is also used to balance the thrusts from the arches. Since the deck has sufficient axial strength to carry the tension resulting from the arch thrusts no other supports are needed to balance the horizontal forces from the arches. The idea for using the bridge deck to balance the arch thrusts appeared to come out of solving the problem that no external horizontal supports need to be provided for a lift bridge.

A similar example, shown in Figure 4.27, is an arch supported bridge for trams in Manchester, England. The integrated arch-cable-deck system achieved self-balancing of horizontal forces, and was a solution of the problem that the site did not allow for building supports to balance the horizontal forces from the arches.

Figure 4.26 Salford Quays Lift Bridge.

Figure 4.27 The arch, deck and cables of the bridge form a self-balancing system.

4.4 Further Comments

The efficiencies of a beam with overhangs and a Y shaped column with a tie at its two top ends have been examined independently in Sections 4.2.1 and 4.2.2, and their implementations have been demonstrated in Sections 4.3.1 and 4.3.2 respectively. However, it is possible and effective to integrate the two physical measures into one design simultaneously. The Chengdu East Railway Station, which is one of the largest railway hubs in China and the largest in the West region of the country, is such an example. The station building was constructed in 2011 [4.6].

Figure 4.28a shows the front view of the railway station in Chengdu in which the overhang roof and Y shaped columns can be seen. Looking at one of the steel Y shaped roof supports (Figure 4.28a), the lower part of the Y shaped column is split into two inclined members that are linked by a metal piece with three pairs of short, horizontal members. The two branch members of

(a)

(b)

Figure 4.28 Chengdu East Railway Station. (a) Front view showing the overhanging roof with Y column supports. (b) One of the four roof supports showing the Y column supports in two perpendicular directions (Courtesy of Professor Yuan Feng, China Southwest Architectural Design & Research Institute, China).

the Y shaped column evolve into pairs of loops. Figure 4.28b shows one of the four roof supports, which can be seen as a Y column being split into four looped branches that are linked by horizontally parallel members. The widely spread branches provide four point supports to the roof structure. As the four top ends are connected to the roof structure, their deflections in the two horizontal directions are constrained, which also limits the bending moments in the members of the Y shaped column.

In this chapter, only vertical loads have been considered. In reality, lateral loads are of the same importance as the vertical loads, and the action of the lateral loads will be discussed in Chapter 6.

References

4.1   Craig, R. R. *Mechanics of Materials*, John Wiley & Sons, USA, 1996.
4.2   Bennett, D. *Skyscrapers: Form and Function*, Simon & Schuster, New York, USA, 1995.
4.3   Parkyn, N. *The Seventy Architectural Wonders of Our World*, Thames & Hudson, London, 2002.
4.4   Xu, L. *The Y-Shaped Structures*, MSc Dissertation, The University of Manchester, 2015.
4.5   Torroja, E. *The Structures of Eduardo Torroja: An Autobiography of an Engineering Accomplishment*, F W Dodge Corporation, USA, 1958.
4.6   Feng, Y. *et al. Practice on Long-Span Spatial Structures*, China Construction Press, Beijing, China, 2015.

Chapter 5

More Uniform Distribution of Internal Forces

5.1 Routes to Implementation

Achieving more uniform distribution of internal forces will lead to smaller internal forces. Therefore, the routes to implementing the structural concept of smaller internal forces presented in Section 4.1, such as using self-balancing, internal and external elastic supports, *etc.*, are all applicable to realising more uniform distributions of internal forces. However, creating more uniform distributions of internal forces provides an alternative way of thinking and can lead to a topology optimisation method for achieving more efficient structures.

Topology optimisation of structures: Evolutionary Structural Optimisation (ESO) and its later development, bi-directional Evolutionary Structural Optimisation (BESO) [5.1, 5.2], are a kind of topology optimisation method in which a structural concept is embedded. By gradually removing inefficient materials with the lowest stresses from a structure and adding material to the most stress demanding region, an optimum topology of the structure evolves with the remaining elements having a smaller difference between the highest and lowest stresses. The maximum stress difference between remaining elements of the structure gradually becomes smaller through repeating this process. The outcome from BESO is an efficient design that wisely uses material. This optimisation process corresponds to the structural concept: *the more uniform the distribution of internal forces or stresses, the more efficient the structure.*

Using BESO based on the finite element method implemented on a computer will lead to creative solutions for a wide range of structures, some of which can be imaginative and even beyond what an experienced engineer could think of.

5.2 Hand Calculation Examples

5.2.1 A Cantilever with and without an External Elastic Support

This example shows that the provision of an external elastic support reduces the bending moments or makes more uniform distributions of bending moment and leads to smaller deflections.

Figure 5.1 shows three vertical cantilevers that have the same height of L and same cross-sectional rigidity of EI and are subjected to a uniformly distributed lateral load of q. The difference between the first two cantilevers is that Cantilever 1 is an unrestrained cantilever and Cantilever 2 is a cantilever with a horizontal spring support at its top end. The spring for Cantilever 2 has the stiffness of K_x. The differences between the Cantilever 2 and Cantilever 3, which also has a spring support at its top end, are that the load and the spring have angles ϕ and θ to the cantilever. Calculate and compare the bending moments at the bases and the deflections at the tops of Cantilevers 1 and 2.

For Cantilever 1, the maximum bending moment at the base and the maximum deflection at the free end of the cantilever are respectively [5.3]:

$$M_{1b} = \frac{q_x L^2}{2} \tag{5.1}$$

$$\Delta_1 = \frac{q_x L^4}{8EI} \tag{5.2}$$

For Cantilever 2, which is a statically indeterminate structure, the spring force needs to be determined before calculating the bending moment and deflection of the cantilever. The spring action can be replaced by a spring force, F_x, to be determined, which can be expressed as a product of the stiffness K_x and the deflection Δ_2 of the spring. Δ_2 is the summation of two deflections, Δ_{2q} and Δ_{2s}, induced by the distributed load and by the spring force on the statically determinate cantilever respectively:

$$F_x = k_x \Delta_2 = k_x (\Delta_{2q} - \Delta_{2s}) = k_x \left(\frac{q_x L^4}{8EI} - \frac{F_x L^3}{3EI} \right) \tag{5.3}$$

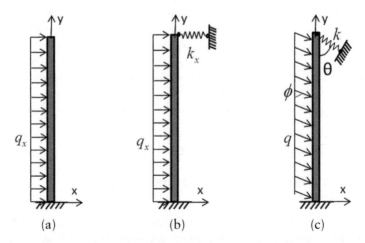

Figure 5.1 A cantilever subjected to uniformly distributed loads. (a) Cantilever 1. (b) Cantilever 2: with an external spring support at its top end. (c) Cantilever 3: the load and the spring support are inclined to the cantilever.

Re-rearranging equation 5.3, the expression of the spring force F_x can be obtained, step by step, as:

$$F_x = \frac{k_x \dfrac{q_x L^4}{8EI}}{1 + \dfrac{k_x L^3}{3EI}} = \frac{3q_x L}{8} \frac{k_x \dfrac{L^3}{3EI}}{1 + \dfrac{k_x L^3}{3EI}} = \frac{3q_x L}{8} \frac{k_x}{K_{sc} + k_x} = \frac{3q_x L}{8} \frac{\alpha}{1 + \alpha} \tag{5.4}$$

in which:

$$k_{sc} = \frac{3EI}{L^3}; \quad \alpha = \frac{k_x}{K_{sc}} \tag{5.5 a, b}$$

where K_{sc} is the static stiffness of the cantilever and is the inverse of the displacement at the top of the cantilever due to a unit load, and α is the ratio of the spring stiffness to the static stiffness of the cantilever. It can be observed from equations 5.4 and 5.5 that:

- When the spring has an infinitive stiffness, $k_x = \infty$, it becomes a roller support and then $F_x = 3q_x L / 8$. This is just the reaction force of a propped cantilever at the roller support.
- The spring force F_x depends on the ratio of the spring stiffness to the static stiffness of the cantilever.

With the spring force known, the cantilever with a spring support becomes a statically determinate structure and the bending moment and deflection at any point of the beam can be easily calculated. For illustration, consider the bending moment at the base and the deflection at the top of the cantilever. The superposition method can be used for calculation:

$$M_{2b} = \frac{1}{2} q_x L^2 - FL = \frac{1}{2} q_x L^2 - \frac{3}{8} q_x L^2 \frac{\alpha}{1 + \alpha}$$
$$= \frac{1}{2} q_x L^2 [1 - \frac{3}{4} \frac{\alpha}{1 + \alpha}] = \frac{1}{2} q_x L^2 \times f_M \tag{5.6}$$

$$\Delta_2 = \Delta_{2q} - \Delta_{2F} = \frac{q_x L^4}{8EI} - [\frac{3q_x L}{8} \frac{\alpha}{1 + \alpha}] \frac{L^3}{3EI}$$
$$= \frac{q_x L^4}{8EI} (1 - \frac{\alpha}{1 + \alpha}) = \frac{q_x L^4}{8EI} \frac{1}{1 + \alpha} = \frac{q_x L^4}{8EI} \times f_\Delta \tag{5.7}$$

$$f_M = 1 - \frac{3}{4} \frac{\alpha}{1 + \alpha} = 1 - \frac{3}{4} \frac{K_x}{K_{sc} + K_x} \tag{5.8}$$

$$f_\Delta = \frac{1}{1 + \alpha} = \frac{K_{sc}}{K_{sc} + K_x} = 1 - \frac{K_x}{K_{sc} + K_x} \tag{5.9}$$

where f_M and f_Δ are the spring effect factors for the base bending moment and for the top deflection of the cantilever respectively, which describes what reductions are achieved due to the spring effect. It can be observed from equations 5.6–5.9 that:

- When $k_x = \infty$, it becomes a propped cantilever. Thus, $M_{2b} = q_x L^2 / 8$ and there is no deflection at the propped position.
- When $k_x = 0$, it becomes a cantilever and the bending moment at the base is $M_{2b} = M_{1b} = q_x L^2 / 2$ and the deflection at the top is $\Delta_2 = \Delta_1 = q_x L^4 / (8EI)$
- When K_x is between the two extremes, the larger the spring stiffness, the larger the spring force and thus the smaller the bending moment at the base and the smaller the deflection at the top of the cantilever.

Equations 5.8 and 5.9 indicate that f_M and f_Δ are functions of the ratio of the spring stiffness to the static stiffness of the cantilever. To appreciate the effect of the stiffness ratio on the reduction of the responses, these two functions are plotted in Figure 5.2.

Figure 5.2 indicates that the spring can effectively reduce the base bending moment and the top deflection of a cantilever, and the rate of the reduction of the bending moment at the base becomes small when α is larger than 3.

For Cantilever 3, which is a development of Cantilever 2 in Figure 5.1b obtained by inclining the spring support with an angle of θ and the distributed load with angle of ϕ. When $\theta = 90^0$, and $\phi = 90^0$, Cantilever 3 becomes Cantilever 2. Equations 5.3–5.8 are applicable to Cantilever 3 by introducing $k_x = k \sin^2\theta$, $F_x = F \sin\theta$ and $q_x = q \sin\phi$, where k_x is the spring stiffness k projected to the horizontal (x) direction. F is the spring force and F_x is the horizontal projection of F. Similarly, q_x is perpendicular to the cantilever, which is the projection of the load q to the x direction. The derivation of k_x can be seen in the example in Section 6.2.1, in which inclined strings, similar to the spring, are used.

5.2.2 An Eight Storey, Four Bay Frame with Different Bracing Arrangements

This example shows that the computer application of the structural concept of the uniform distribution of internal forces can generate new bracing patterns that are even more efficient than those based on the structural concept of direct internal force path.

Figure 5.3 shows an eight storey, four bay frame with five different bracing arrangements. Frames A-D have the same dimensions, the same numbers of bracing, vertical and horizontal members. All the members have the same cross-sectional area and elastic modulus. Two bracing members are placed on each of the eight storeys in Frames A-D, making a total 16 bracing members. The only differences between the four frames are the bracing patterns. The bracing pattern in Frame A has been discussed in Chapter 3 and can be formed by using the first three criteria in Section 3.1. The inverted V pattern in Frame B is

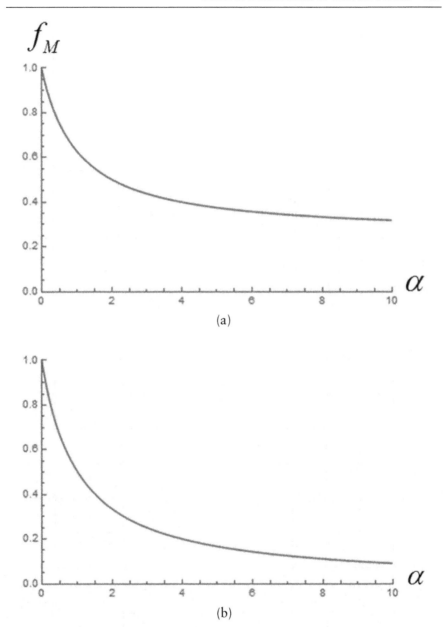

Figure 5.2 Spring effect factors as functions of the stiffness ratio. (a) For the base bending moment. (b) For the deflection at the top of the cantilever.

developed intuitively based on the structural concept of smaller internal forces. Frames C and D are developed using the Evolutionary Structural Optimisation for continuous bodies by applying it to discrete systems. Frame E is initially fully braced so that each panel has two braces. A pair of anti-symmetric loads is

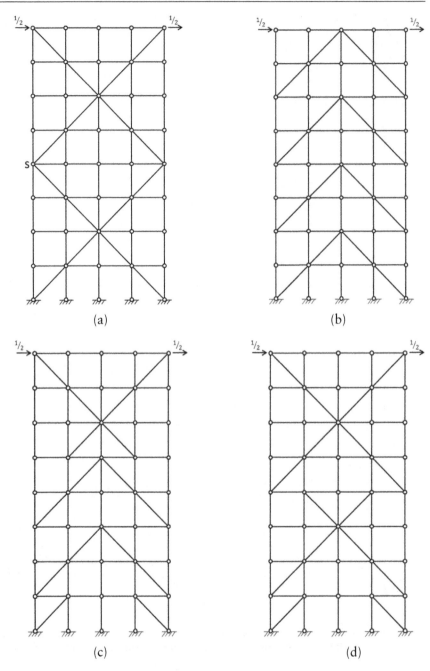

Figure 5.3 A frame with four different bracing arrangements. (a) Frame A: Global X bracing derived from the structural concept of direct force paths. (b) Frame B: Inverted V bracing derived from the structural concept of smaller internal forces. (c) and (d) Frames C and D with bracing patterns derived from the ESO approach based on the structural concept of more uniform distribution of internal forces. (e) Frame E: Fully braced frame as a start for the ESO approach.

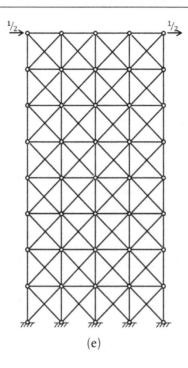

(e)

Figure 5.3 (Continued)

applied to the two top corner nodes as shown in Figure 5.3. Finite element structural analysis of the pin-connected frame is conducted and the pair of symmetric bracing members that have the smallest strain energy (or stress) are identified and removed from the structure. This process is repeated until only two bracing members are left in each storey. Frames C and D are the outcomes from ESO. Due to the removal of the bracing members with the lowest stain energy, the two evolved structures have *smaller differences in strain energy* between the remaining members, which is equivalent to a realisation of *more uniform distribution of internal forces* [5.4]. Using Frames A-D shown in Figure 5.3, the internal forces and the maximum deflections of the four frames can be examined.

The pin-connected frames are statically indeterminate structures. However, they can be simplified into equivalent statically determined structures following the concept: *when a symmetric structure is subjected to anti-symmetric loads, the responses (deflections and internal forces) of the structure will be antisymmetric*, which has been used in Chapters 2 and 3. As the responses are anti-symmetric, the members in the central line of the frame must be zero and there are no vertical deflections of the nodes on this line. Therefore, the equivalent half frames are shown in Figure 5.4 in which the internal forces of all members are indicated to appreciate their magnitudes and distributions. The internal forces of all members of the four halved frames can be calculated by hand using the equilibrium equations at each of the nodes. As the length and width of each panel are the same, the hand calculation can be quickly conducted.

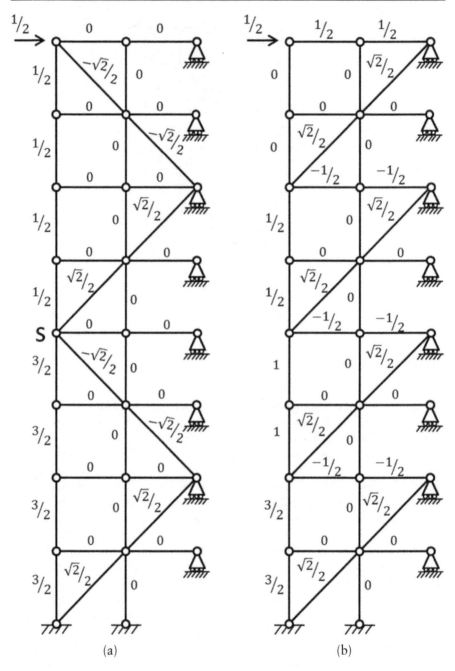

Figure 5.4 The internal forces for the four equivalent half frames based on those shown in Figure 5.3. (a) Frame A equivalent. (b) Frame B equivalent. (c) Frame C equivalent. (d) Frame D equivalent.

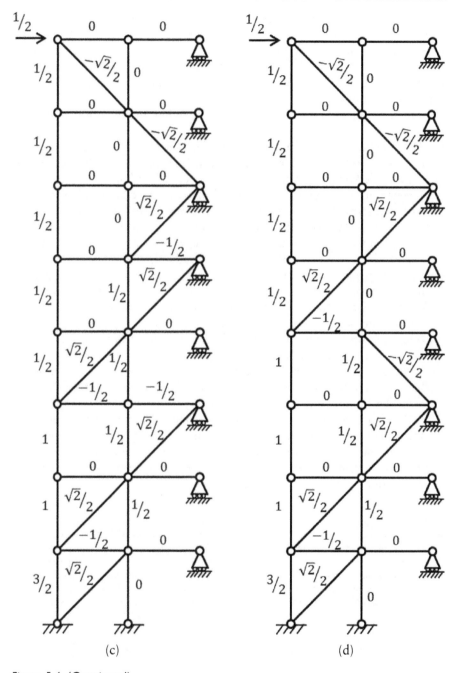

Figure 5.4 (Continued)

To understand the differences of the internal forces between the four frames, Table 5.1 summarises the distribution and magnitude of the internal forces in the vertical and horizontal members (columns 2–5) and in the bracing members (column 6) based on Figure 5.4 and their contributions to the deflection (columns 7 and 8) and the total contributions (column 9) of the four halved frames.

It can be noted and understood from Figure 5.4 and Table 5.1 that:

- The bracing members have the same magnitude of $\left|\sqrt{2}\,/\,2\right|$ in all four frames. Therefore, the difference between the deflections of the four frames is controlled by the internal forces in the vertical and horizontal members.
- The bracing arrangement in Frame A, following the concept of more direct path of internal forces, has the highest number (24) of zero-force members but gives the largest deflection among the four. This is because there are four members with the largest internal force of 3/2 (Figure 5.4a), which makes a significant contribution to the total deflection.
- In comparison with that of Frame A, the inversed V-shape bracing pattern have six more members with a small internal force of 1/2 and two more members with a force of 1 but two less members with the largest magnitude of internal force of 3/2. Due to the square effect on the forces in equation 2.16, this bracing pattern results in a smaller deflection than that of Frame A with a global X bracing pattern.
- In comparison with Frame A and B, Frames C has higher numbers of members with non-zero force and with small magnitude of internal force (1/2) but a smaller number of members with the largest magnitude of internal force (3/2). Therefore, Frame C results in even smaller deflection than Frame B.
- Frame D has the same deflection as Frame C by reducing four members with an internal force of 1/2 but increasing one member with an internal force of 1. The two sets of members have the same contributions to the deflection.

Table 5.1 The number of members at different magnitudes of internal forces and their contributions to deflections of the four frames

(1)	Internal forces (IF) in the vertical (V) and horizontal (H) members				IF in the Bracing members	Contributions from V and H members	Contributions from bracing members	Contributions From all members
	(2)	(3)	(4)	(5)	(6)	(7)	(8)	(9)
	0	½	1	3/2	$\sqrt{2}/2$	$\overset{V,H}{\sum} N^2 L$	$\overset{B}{\sum} N^2 L$	$\overset{All}{\sum} N^2 L$
Frame A	24	4	0	4	8	10	$4\sqrt{2}$	15.65
Frame B	18	10	2	2	8	9	$4\sqrt{2}$	14.65
Frame C	16	13	2	1	8	7.5	$4\sqrt{2}$	13.16
Frame D	19	9	3	1	8	7.5	$4\sqrt{2}$	13.16

When a bracing member ends to a vertical member, the internal force in the vertical member that is lower than the intersection point will increase by 1/2. For example, there are two bracing members that end at the intersection point (S) of the outside vertical members at the mid-height of Frame A in Figures 5.3a and 5.4a. Therefore, the internal force increases $2 \times 1/2 = 1$ from the upper member of the intersection point to the lower member. To avoid larger accumulated internal forces in the outside vertical members, two bracing members end at two inside columns next to the outside ones at the six level of Frame C in Figure 5.3c. This leads Frame C to more members having smaller internal forces and fewer members experiencing larger internal forces in comparison to Frame A (Figures 5.4a and 5.4c).

The four bracing patterns are generated based on different structural concepts, the more direct internal force path (Frame A), the smaller internal forces (Frame B) and more uniform distribution of internal forces (Frames C and D). Therefore, the outcomes shown in Table 5.1 encourage to think retrospectively and conceptually the reasons that the frame with the last three bracing patterns (Figures 5.4b, c, and d) perform even better than the globally X braced frame (Frame A), which helps to develop ideas for wiser designs. On the other hand, ESO would be able to create new structural forms that may be beyond what one can image for.

5.3 Practical Examples

5.3.1 Structures with External Elastic Supports

5.3.1.1 Samuel Beckett Bridge, Dublin

A simple harp, shown in Figure 5.5a, consists of three relatively thick external members, the neck with harmonic curve, a sound box and a pillar, which form a loop, and a number of parallel taut strings with different lengths between the neck and the sound box. The function of the pillar is to support the neck, prevent relative deformation between the neck and the sound box due to the action of the taut strings, and transmit its supporting force from the neck to its lower end and to the base.

For analysing the forces in a harp, a hand drawing of the essence of a harp similar to that in Figure 5.5a is shown in Figure 5.5b. If the pillar is removed from the harp, its action forces on the neck and the sound box and the remaining structure are as shown in Figure 5.5(c). The downward force, F, acting on the sound box would be transmitted to the base that supports the sound box. Therefore, supports are qualitatively provided to allow the remaining parts of the harp in an equilibrium position. The upward compressive force, F, is necessary to support the neck. From a structural point of view, this upward compressive force can be replaced by a suitable tension force, T, as shown in Figure 5.5(d), to resist the downward deformation of the neck.

The idea of the harp as illustrated qualitatively in Figure 5.5(d) has been used in engineering practice in the Samuel Beckett Bridge, which is a cable-stayed, steel box girder, swing bridge over the River Liffey in Dublin. This bridge, designed by Santiago Calatrava, and has become a landmark of the city reflecting a harp which is the national symbol of Ireland. The side view of the bridge shown in Figure 5.6a looks like the equivalent harp shown in Figure 5.5d in which the cables between the bridge deck and the pylon resemble strings, the pylon acts like the neck of the harp and the bridge deck is similar to the music box. The backstay cables provide tension forces to limit the forward and downward deformations of the pylon due to the action of the main cables.

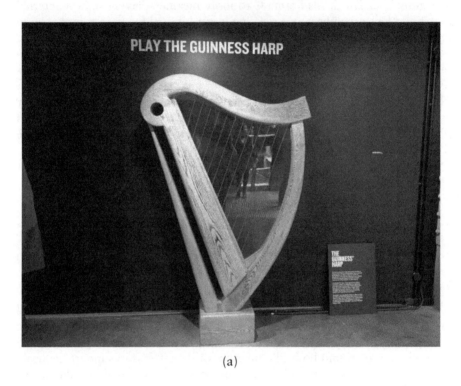

(a)

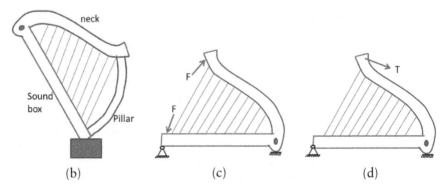

(b)   (c)   (d)

Figure 5.5 A harp. (a) The harp displayed at the Guinness factory, Dublin. (b) Hand drawing of a harp similar to that in (a). (c) Replacing the pillar with two forces. (d) The upward compressive force *F* is replaced by a tension force *T*.

(a)

(b)

Figure 5.6 The Samuel Beckett Bridge, Dublin. (a) Side view. (b) Back view.

The bridge is 123m in length and 28m in width and carries four lanes of traffic with cantilevered pedestrian and cycle tracks. The bridge is able to rotate through 90 degrees to allow shipping to pass below. Figure 5.6 presents the side and back views of the bridge, which shows the structure of the bridge and its use. The steel box girder bridge is lifted by 25 stay cables of 60mm diameter from a steel cantilever that is supported and stabilised by six back cables of 145mm diameter. The positions of the six back cables also strengthen the lateral stability of the pylon [5.5].

Figure 5.1c also suggests a plane model for the analysis of the pylon. The pylon is subjected to a series of parallel cable forces that are about perpendicular to the pylon while the back-stay cables between the top of the pylon and their anchor points act like spring supports to constrain the downward and forward deflections of the pylon. In this case the back cables are anchored to the foundations that are independent from the bridge and thus can be considered to act as an external spring support to the pylon.

5.3.1.2 Serreria Bridge, Valencia

There are other harp-like bridges which have been designed by Santiago Calatrava. The Serreria Bridge, shown in Figure 5.7, is situated within the City of Arts and Science Complex in Valencia, Spain. It has a span of 180m and a width varying from 33.5m to 39.2m. In addition to three external vertical supports, the bridge deck is suspended by 29 parallel stay cables with a spacing of 5m from an inclined curved pylon that rises to a height of 118.6m. It can be observed from Figure 5.7 that the pylon leans backwards which enables the self-weight of the pylon to balance some of the applied forces from the stay cables supporting the bridge deck. The two groups of back-stay cables provide effective external spring support to the pylon at its top and are placed slightly more apart from each other on ground to improve the lateral stability of the pylon.

Figure 5.7 The Serreria Bridge, Valencia, Spain (Courtesy of Mr. Nicolas Janberg, structurae.net, Germany).

5.3.1.3 Katehaki Pedestrian Bridge, Athens

The Katehaki pedestrian bridge is 93.7m long and spans 73.5m between two supports. The width of the deck varies from 3.95m at one end to 5.67m at the other end. The curved steel-boxed pylon has a height of 50.48m and appears to lean backwards, while the pylon in the Samuel Beckett Bridge leans forward. The bridge deck is suspended from the pylon by 14 parallel cables. Two back-stay cables are able to transmit a significant portion of the parallel cable forces to their foundations. The bending moments in the pylon caused by the back-stay cables offset part of the bending moments due to the parallel cable forces.

The three bridges are functionally, geographically and architecturally different, but they demonstrate similar technical elegance and lightness, achieved by providing external elastic supports:

1. In addition to the solid vertical supports, the bridge decks are suspended by a series of cables that act as external spring supports to the decks.
2. The back-stay cables provide external spring supports to the pylons that act like cantilevers. The effect of such spring supports on a cantilever has been demonstrated in Section 5.2.1

For a quick hand analysis at a conceptual design stage, the pylons in the three bridges can be considered as cantilevers with spring supports at their free ends (Figure 5.1c). The forces acting on the pylons from the parallel fore-stay cables linking to the bridge decks may be treated as uniformly distributed loads, which may not be perpendicular to the pylons and can be described by an angle ϕ. The back-stay cables can be simplified as an external spring support at the free end of the pylon, and θ is used to define the angle between the cables and the pylon.

5.3.2 Structures with Internal Horizontal Elastic Supports

The provision of internal elastic supports is perhaps an effective and simple way to self-balance internal forces which will result in smaller internal forces and more uniform distribution of internal forces, and smaller deflections. There are several creative applications of internal elastic supports in the horizontal direction.

5.3.2.1 Manchester Central Convention Complex, UK

Figure 5.9 shows the Manchester Central Convention Complex that has a distinctive arched roof with a span of 64m. The Complex was originally designed in 1880 and subsequently used as the Manchester Central Railway Station. The roof arches were made of wrought iron. Arches are effective structures as they transfer applied loads mainly through compression, rather than by bending, to their supports. However, arches normally generate large horizontal forces

Figure 5.8 Katehaki pedestrian bridge, Athens.

Figure 5.9 Front view of the Manchester Central Convention Complex (MCCC), Manchester.

at supports, which require large foundations. Normally pinned supports are provided at the two ends of an arch to resist both vertical and horizontal forces. It can be observed on the arch shown in Figure 5.9 that there are two substantial horizontal members, one toward the bottom of the arch and one around mid-height of the arch. The self-weights of the two horizontal members are transmitted through the vertical bars to the arch. The two horizontal members have large axial stiffnesses and effectively act as internal spring supports to the arch in the lateral direction, which restrains lateral deformations of the arch and balances part of the horizontal component of the internal forces in the arch. This in turn reduces the internal forces in the arches and reduces the horizontal thrusts at the arch supports.

To examine the effects of the horizontal members on the reduction of internal forces and deflections in an arch in both horizontal and vertical directions qualitatively, the main characteristics of the arch (Figure 5.9) can be extracted as shown in the simplified model (Model A) in Figure 5.10a, in which the two horizontal members and the boundary conditions of the arch are shown. It may not be an easy task to produce a physical model like that in Figure 5.10a because supports and connections between the supports and the arch are concerned. Using the concept of symmetry (when a symmetric structure is subjected to symmetric loads, the response of the structure will be symmetric), Model A (Figure 5.10a) is just a half of Model B (Figure 5.10b). The advantage to use Model B to replace Model A is that the supports required in Model A can be removed for model making. For examining the effect of the horizontal members on the ring, Model

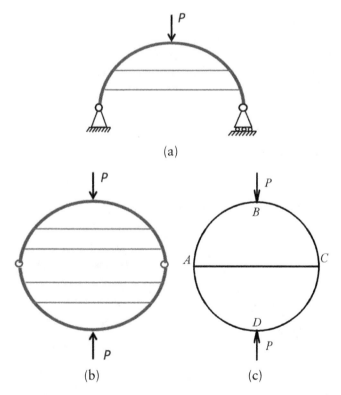

Figure 5.10 Evolution of models for intuitive understanding. (a) Model A: the arch-bar model based on MCCC. (b) Model B: The model is equivalent to Model A based on symmetry. (c) Model C: This model is a simplification of Model B, which captures the physical essence of Model B.

B can be altered to Model C (Figure 5.10c) in which the physical essences of the horizontal members and the arch remain. The changes between Models B and C are the replacement of the four horizontal members by a single member and the replacement of the pin connections by rigid connections. The behaviour of Model C is similar to that of either Model A or Model B, and it is easier to make a physical model of Model C than Models A and B.

Due to the action of the vertical forces, P, the horizontal member in Model C experiences a tensile force T. The behaviour of the ring without a tendon but still subjected to forces P and T can be analysed qualitatively using the super-position method. Considering the pair of vertical forces P alone, the deformation of the ring is illustrated by the dashed line in Figure 5.11a, in which the top and bottom points B and D deform toward to each other while the side points A and C deform outwards from each other. Examining the action of the horizontal forces T, the ring deforms in the opposite direction to that induced by P (Figure 5.11b). Therefore, the behaviour of Model C is a combination of the deformations shown in Figures 5.11a and 5.11b. The action of the horizontal member confines the outward deformation of A and C and reduces the vertical deflection of the ring due to P. The tied ring is much stiffer than the corresponding ring without a tie. This interpretation can be demonstrated using physical models.

Figure 5.12 shows two rubber rings, one with and one without a wire tied across the centre of the ring. The same weight of 22.3N is placed on the top of each of the two rings, and the reduced deformation of the tied ring is apparent. The reduced deflection means an increased stiffness of the ring

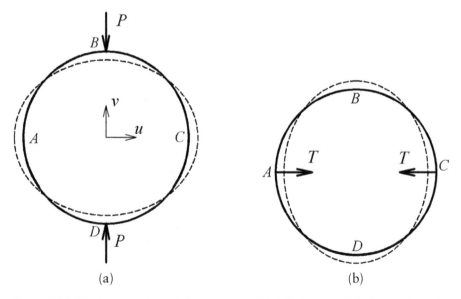

(a)   (b)

Figure 5.11 Illustration of the deformations of the tied ring. (a) A ring subjected a pair of vertical forces and its deformations. (b) A ring subjected to a pair of horizontal forces and its deformations.

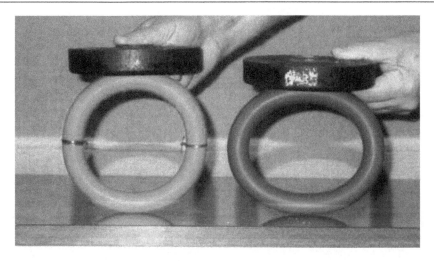

Figure 5.12 Demonstration of the effect of the wire that acts as internal elastic supports to the ring.

which can be felt easily by pressing down on the two rings. The reduced deflection also indicates that the tied ring experiences smaller bending moments. This is because the force in the wire increases as the applied load increases and produces a bending moment in the ring in the opposite direction to the bending moment caused by the external load. Thus, the force in the wire balances part of the bending moments in the ring due to the vertical load, resulting in smaller and more uniform internal forces. As the tied ring is doubly symmetric and relatively simple, the expressions for its vertical and lateral deformations and bending moments can be derived and quantitatively examined [5.6].

5.3.2.2 Raleigh Arena, USA

The roof structure of the Raleigh Arena shown in Figure 5.13 consists of carrying (sagging) cables and stabilising (hogging) cables, which are supported by a pair of inclined arches. The carrying cables apply large forces to the arches and some of the vertical components of these forces are transmitted to external columns. The bending moments, shear forces and compressive forces are transmitted through the inclined arches to their supports. Most of the horizontal forces at the ends of the arches are balanced by underground ties which reduce significantly the horizontal forces on the foundations. The underground ties or tendons have a similar function to the wire tie in the rings demonstrated in Figure 5.12, acting as internal elastic supports, reducing the internal forces of the arches and making the structure stiffer.

It is useful to interpret the behaviour of the underground ties further. The lower part of the arches and the tendon, circled in Figure 5.13b, can be simplified as a rigid frame with a tendon linking the two ends of the frame, which is

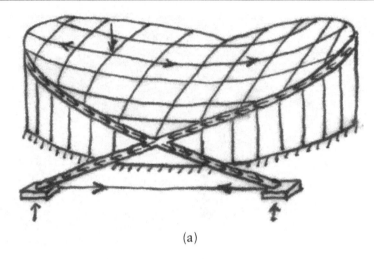

(a)

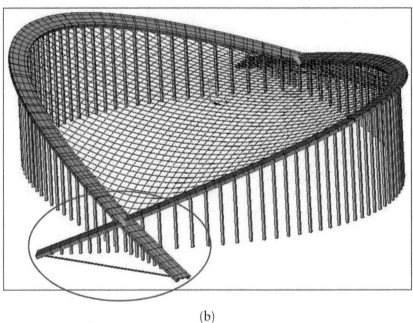

(b)

Figure 5.13 Raleigh Arena. (a) The internal force paths (Reproduced from [5.7]).
(b) The finite element model [5.4].

named Model D as shown in Figure 5.14a. Due to the symmetric nature of the structure, only a vertical force is applied at the top of the model. This vertical load is transmitted through the two inclined members in bending, compression and shear to the tendon and the foundations. Model D can be represented by Model E based on the principle of symmetry (Figure 5.14b). Effectively, the tendons in Models E and C have the same functions, constraining the lateral

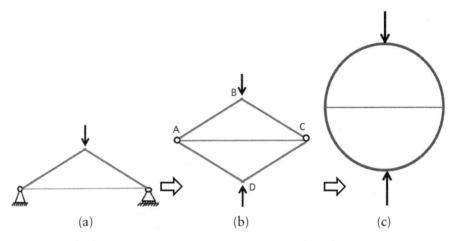

Figure 5.14 Illustration models. (a) Model D: A simplified model to represent the lower part of the two inclined arches for studying the effect of the tendon. (b) Model E: The equivalent model to Model D. (c) Model C: An alternative representation of Model E for studying the effect of the tendon.

relative deflection between nodes A and C and the inward vertical deflection between B and D and reducing the bending moments in the straight and curved members. The tendon action in the structure (Figure 5.13b) can be explained and demonstrated as shown in Figures 5.11 and 5.12.

5.3.3 Structures Derived from Topology Optimisation

5.3.3.1 Evolutionary Structural Optimisation (ESO)

Evolutionary Structural Optimisation (ESO) is a popular and relatively simple topology optimisation method that can be integrated with many commercial finite element analysis software packages. This circumvents the difficulties in solving complicated structural analysis problems. The basic ESO method has been improved and extended to Bi-directional Evolutionary Structural Optimisation (BESO) which allows adding material to the most demanding parts of a structure to enhance structural performance and reduce stress intensity, in parallel with removing material from parts of the structure where it is used to its least advantage. BESO, now a mature technique, is particularly appealing to practising structural engineers and architects because it is well suited to building structures and because a structural concept is embedded into the method.

Adding material to or removing material from a structure in BESO makes the highest stress smaller and the lowest stress larger, which serves to create a more uniform distribution of stress or internal forces in a structure. After removing the material with the lowest stresses from a structure, the lowest stress will become larger while after adding material where stresses are highest, the highest stress will become smaller. In other words, the difference between the maximum and minimum stresses will become smaller than in the original structure. Repeating this process, differences become even smaller and the stress distribution in members becomes more uniform. In seeking more uniform stress-distribution, BESO can create new structural topology. The structural concept embedded in the method can be expressed as ***the more uniform the distribution of stress, the more efficient the structure,*** in which efficiency is measured by the uniformity of the stress distribution in the structure. This structural concept is similar to the one studied earlier in this chapter, ***the more uniform the distribution of internal forces, the smaller the deflection.*** The BESO process, which is an automatic computational process, applies to local areas of a structure following the solution of equilibrium equations and gradually evolves the original structure into a new structure with a superior geometrical form which is usually structurally efficient and aesthetically pleasing.

The topology optimisation problem in BESO is presented as follows [5.1, 5.2]:

Find X, so that:

$$\text{Minimise } C = \frac{1}{2} P^T U = \frac{1}{2} \sum P_i u_i \tag{5.1}$$

$$\text{Subjected to } KU = P \tag{5.2}$$

$$X^T V = V^* \tag{5.3}$$

where X is the design variable vector in which $xi/$(the ith element in the X vector) takes either 0 for the relevant element being absent or 1 for it being present; P and U are the external load vector and the nodal displacement vector respectively, and C is the objective function and is called the mean compliance that indicates the averaged structural flexibility. In other words, C is the inverse of the overall stiffness of a structure. C is the same as W_{11}, the work done by the external loads P on the corresponding deflections U in equation 2.9. Equation 5.2 is the equation of equilibrium. Equation 5.3 is the constraint condition that the prescribed volume limit of the whole structure, V^*, equals the sum of the element volumes in which v_i is the element volume.

There are similarities and differences between the BESO method and the method using the structural concepts proposed in this book. Table 5.2 summarises the main features of the two methods for achieving more efficient structures.

Further comments on the comparison of the two methods for design are as follows:

1. **The objective**
    -   The minimum averaged structural flexibility for a given body mass is searched for BESO while the smaller deflection $\Delta_{2,C}$ (equation 2.16), or

Table 5.2 Comparison between BESO Method and SCM

	BESO method [5.1, 5.2]	Structural Concepts Method (SCM)
Objective	Minimising the objective function, *i.e.* the averaged structural flexibility.	Making the deflection at the critical point of a structure smaller.
Design variables	Elements being present or absent.	Internal forces.
Constraints	Equation of equilibrium and the prescribed volume of the structure.	Not explicitly presented.
Intermediate outcome	N/A	The four structural concepts.
Solution process	Computer process by gradually removing less effective material and adding material to the most needed region.	Using the structural concepts to improve the flow or distribution of internal forces.
Final outcome	A new geometry of the structure with the minimum averaged structural deflection for a given body volume.	A structure with a smaller deflection in comparison with similar structures.

the smaller maximum flexibility coefficient in the flexibility matrix of a structure, is pursued qualitatively for SCM.
- The actual loads acting on the structure are used with BESO while for SCM the actual loads are lumped and normalised at the critical point of the structure.
- The external work $P^T U / 2$ is explicitly calculated in BESO while internal energy is qualitatively interpreted in SCM.

2. **Design variables:** The design variables are the elements that can be either present or absent in BESO while the design variables are the internal forces in the elements with SCM.

3. **Constraints:** The constraints control how much material is to be removed from the original structure in BESO while there is no similar explicit constraint in SCM.

4. **Intermediate outcome:** The intermediate outcome is not needed as the optimum topology is the final outcome in BESO while for SCM, four structural concepts have been identified from the objective function (equation 2.16), which need to be implemented through developing particular physical measures.

4. **Solution process:** The BESO method is implemented with a finite element analysis package and processed using a computer. Therefore, the user should be familiar with the method and the package. Although no computer is necessarily required for SCM, an experienced structural engineer

or architect is needed to design the flow or distribution of the internal forces using any of the four structural concepts.

5. **Final outcome:** The BESO method creates a new structural geometry that may be very different from the original geometry, perhaps even a topology beyond one that could be imagined. The SCM route is likely to achieve a rational design that has relatively small deflections in comparison to other similar structures.

Three illustration examples of bridge design carried out using BESO [5.8] are now considered, in comparison with similar practical examples.

5.3.3.2 A Bridge with a Flat Deck on the Top

Figure 5.15a shows a uniform block that consists of a non-designable layer on the top surface and a design domain that is to be designed as a bridge. A uniformly distributed load of $100N/m^2$ is applied on the top surface of the block and the four bottom corners are pin supported. Steel is used with the elastic modulus $E = 210GPa$ and Poisson's Ratio $v = 0.3$. In order to create a clearance under the bridge, an artificial constraint is added into the design domain by introducing a void strip under the deck in the middle face.

The constraint of removing 80% of the materials in the design domain was applied when BESO was carried out and the optimal topology solution

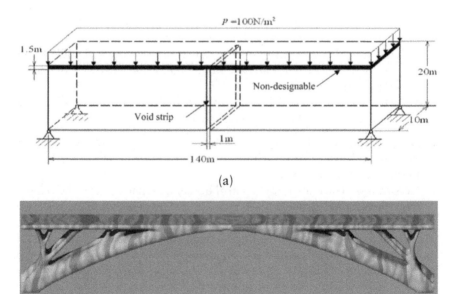

(a)

(b)

Figure 5.15 Comparison between designs from BESO and from structural concepts. (a) The design domain of a top loaded bridge with pinned supports at the bottom corners. (b) The BESO design (Courtesy of Professor Yi Min Xie, RMIT University, Australia).

Figure 5.16 A cast iron arch bridge in Manchester.

produced is shown in Figure 5.15b, which is an arch bridge. As the given boundary conditions can take horizontal forces, the arch design would be an expected outcome. Considering possible designs for the bridge conceptually and intuitively, an arch design would be a likely reasonable solution. Figure 5.16 shows such an arch bridge made by cast iron in Manchester, UK, which is globally similar to the BESO design in Figure 5.16b. The differences between the two designs are the locations and orientations of the columns between the arch and the deck. In a structural concepts design, vertical members with equal spaces are likely to be used, as is usually the case in practice. The inclined deck support members generated in the BESO design are less likely choices, but it shows a more ideal design that leads to a more uniform distribution of stress.

5.3.3.3 A Bridge with a Flat Deck at the Middle Level

Figure 5.17a shows a H shaped uniform block with the central horizontal layer being designated a non-designable layer and four corner pin supports. A uniformly distributed load is applied on the horizontal layer. The two vertical elements of the H section are the design domain and materials can be removed from this domain. The constraint is set to remove 90% of the material from the design domain in the BESO process.

Figures 5.17b shows a 3D print-out of the optimum solution produced by the BESO process, which is an arch bridge with tension members supporting the central section of the deck and compression members supporting the two ends of the deck. This optimised topology reflects well good engineering practice for an efficient structure. One such example is the Tyne Bridge over the River Tyne, linking Newcastle upon Tyne and Gateshead in the UK, which was completed in 1928, as shown in Figure 5.18. The vertical cables (tension members) support the central section of the bridge deck and vertical props (compression members) support the end sections of the bridge deck. Minor differences between the topologies of this bridge and the BESO design (Figure 5.17b) are

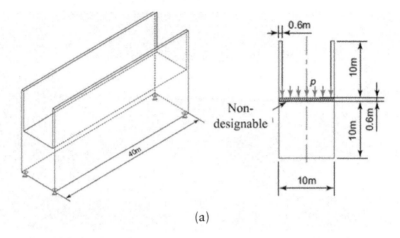

(a)

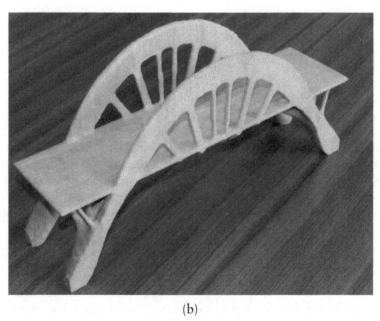

(b)

Figure 5.17 Topology optimisation of a bridge. (a) Design and non-design domains. (b) 3D print-out of the BESO outcome (Courtesy of Professor Yi Min Xie, RMIT University, Australia).

the orientations and cables and the props, which are inclined in the BESO design but are vertical in the actual bridge. It would be interesting to compare the structural performances of the two arrangements for tension members!

5.3.3.4 A Long-Span Footbridge with an Overall Depth Limit

This design requires an arch shape bridge with a 72m clear span between two piers and a maximum arch depth of 1.8m. The BESO process was used to

Figure 5.18 The Tyne Bridge in Newcastle upon Tyne, UK.

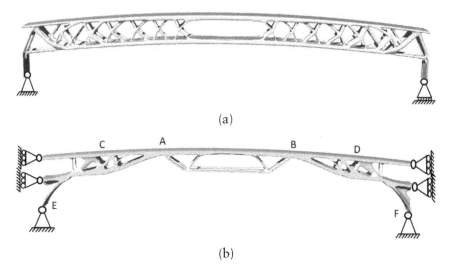

Figure 5.19 The optimised geometries of the bridges. (a) Bridge with pin and roller supports. (b) Bridge with pin supports and side roller supports.

create two structurally efficient and aesthetically pleasing forms for the footbridge. 3D brick elements were employed in a finite element model of the design domain of the bridge and a single material of steel was used. A uniformly distributed load is applied on the top of the structure. Two different boundary conditions were considered: 1) a roller support at the bottom of one pier and a pin support at the other pier; 2) pin supports at both lower ends of the bridge as well as horizontal pin supports at the two ends of the bridge.

The two solutions for the different boundary conditions are shown in Figure 5.19. Since the roller support at one pier allows for horizontal movement, the bridge in Figure 5.19a acts like a simply supported beam for which the behaviour is well known. The bending moments are the largest at the centre of the span where the shear forces are the smallest. The BESO solution shows that materials are only placed at the top and bottom in the central part of the bridge to resist bending moments and material at the top and bottom gradually reduces away from the centre of the bridge, reflecting the variation of the bending moment along the bridge. The inclined members gradually become thicker from the centre to the two ends of the bridge reflecting the variation of shear force.

As much stronger boundary conditions were applied to the second design, the BESO process leads to the more efficient design shown in Figure 5.19b. It is perhaps unlikely that the BESO solution would be anticipated but the rationality of the design can be explained. Referring to Figure 5.19(b), from the distribution of members, the region between A and B appears to be dominated by bending with only small bending moments occurring around positions A and B. The inclined members CE and DF provide vertical support at positions C and D effectively reducing the span of the bridge which of course leads to smaller internal forces and deflections.

A similar example from practice is the Kirchheim Overpass, a road bridge built in 1993 in Germany, which is shown in Figure 5.20a. The rigid frame bridge has a pair of inclined legs that provide support to the bridge deck and effectively shorten the bridge span. The inclined legs experience mainly compressive forces rather than bending effects as the deformations of the legs in both horizontal and vertical directions are confined by the deck and the symmetry of the two inclined legs, which is shown in the bending moment diagram due to a uniformly distributed load in Figure 5.20b. This will also be demonstrated by a hand calculation example in Section 6.2.2.

The rationality of the BESO design can be appreciated through a comparison of the form of the structure and the shape of the bending moment diagram

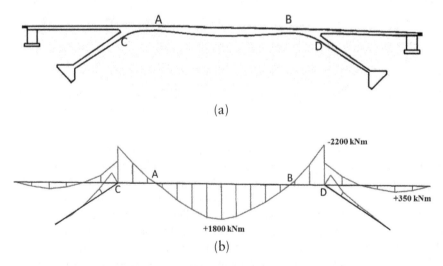

(a)

(b)

Figure 5.20 The structure of the Kirchheim Overpass. (a) Structural form. (b) Bending moment diagram (Reproduced from [5.9]).

for the Kirchheim Overpass bridge. Relatively small bending moments occur at Points A and B in Figure 5.19b, which correspond to the smallest sections of the bridge. Points C and D in Figures 5.19b and 5.20b are where the negative bending moments are the largest and also where the largest bridge sections are (Figure 5.19a).

5.4 Further Comments

The concepts of smaller internal forces and more uniform distribution of internal forces can lead to some similar physical measures for implementation. However, more uniform distribution of internal forces does not necessarily mean smaller internal forces. In BESO, a structure is evolved mainly through removing ineffective materials from the optimisation body. Consequently, the optimised structure would have a more uniform distribution of stress but higher values of stress due to the use of less material.

It is fascinating that a similar structural concept, *the more uniform the distribution of stress, the more efficient the structure*, has been implemented for a computer realisation in BESO. In other words, the structures generated from BESO are likely effective, efficient and possibly elegant. The three BESO bridge examples demonstrate that the BESO process is able to produce good engineering designs, and the optimum topology designs can provide an excellent starting point for practical design. It is of interest that the three comparative practical designs can be evolved from the BESO outcomes with a consideration of practical aspects.

References

5.1   Huang, X. and Xie, Y. M. A Further Review of ESO Type Methods for Topology Optimisation, *Structural and Multidisciplinary Optimisation*, 41, 671–683, 2010.

5.2   Xie, Y. M. and Steven, G. P. A Simple Evolutionary Procedure for Structural Optimisation, *Computer and Structures*, 49, 885–896, 1993.

5.3   Hibbeler, R. C. *Mechanics of Materials*, Sixth Edition, Prentice-Hall Inc., 2005.

5.4   Yu, X. *Improving the Efficiency of Structures Using Mechanics Concepts*, PhD Thesis, The University of Manchester, 2012.

5.5   Olierook, G. *Construction of Samuel Beckett Bridge Dublin—Ireland*, Hollandia, 2009.

5.6   Ji, T., Bell, A. J. and Ellis, B. R. *Understanding and Using Structural Concepts*, CRC Press, Taylor & Francis Group, London, 2016.

5.7   Bobrowski, J. Design Philosophy for Long Spans in Buildings and Bridges, *Structural Engineer*, 64A(1), 5–12, 1986.

5.8   Xie, Y. M., Zuo, Z. H., Huang, X., Black, T. and Felicetti, P. Application of Topology Optimisation Technology to Bridge Design, *Structural Engineering International*, 185–191, 2014.

5.9   Holgate, A. *The Art of Structural Engineering: The Work of Jorg Schlaich and His Team*, Edition Axel Menges, 1996.

Chapter 6

Converting More Bending Moments Into Axial Forces

6.1 Routes to Implementation

1. Using bar/string members to create vertical internal elastic supports

This physical measure follows the route of providing internal elastic supports mentioned in Section 4.1 but makes the route more specific in the vertical direction. It is understood that shortening a span is the most effective way to reduce deflections, but it may not always be feasible due to conflicting structural, architectural or functional requirements. In such cases. Providing vertical internal elastic supports while meeting the other requirements becomes an attractive solution to dealing with deflection.

Beam-string structures, with a variety of forms, have been used as efficient types of structures. The simplest beam-string structure is illustrated in Figure 6.1a.

The basic beam-string structure is a simply supported beam AB with a vertical internal elastic support that is provided by a strut, CD, placed under beam and linked to a profiled string (or tendon), ADB. When the beam deforms downwards due to the action of the load, the strut CD moves down inducing tension forces in string, ADB, which effectively provides an upward force to the beam through the strut, CD. The string and strut act like a spring support to the beam as shown in Figure 6.1b, which converts part of the bending moment in the beam into the axial forces in the strut and string. The effect of the string and strut in the beam-string structure will be examined using a hand calculation example from Section 2.2.1.

2. Using inclined members to replace vertical members.

Columns, as vertical members, have been widely used as supporting elements in frame structures and can be seen in almost every building transmitting vertical loads, mainly through compression, and lateral loads, in unbraced frames, mainly through bending to their foundations.

When a pair of long structural members are inclined to form a triangular frame, they will carry predominantly axial forces for both vertical loads and

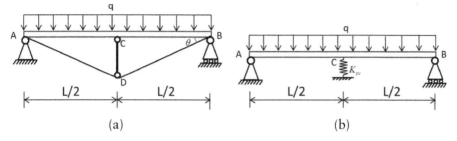

Figure 6.1 Vertical internal spring support. (a) A beam-string structure. (b) An equivalent model for the beam in the beam-string structure.

lateral loads applied to the top of the frame, braced and unbraced. There are a several ways to achieve the ideal situation, two of them are:

a. If the two inclined members lie in a vertical plane and are connected at their top ends, they will be able to resist vertical and lateral loads in the vertical plane and transmit the loads to the supports mainly through axial forces.
b. If the pair of inclined members lean an angle to the vertical plane and is supported by other members for achieving equilibrium, they will be able to resist vertical loads and the lateral loads in the plane and out of the plane mainly through axial forces

3. Orientation of members.

If a cantilever column is orientated away from the vertical, the self-weight of the column will be transmitted through bending to its support. It may be possible to use its self-weight to balance part of the action of the external loads.

6.2 Hand Calculation Examples

6.2.1 A Beam with and without a Vertical Internal Elastic Support

This example shows how the provision of a vertical internal elastic support in a simply supported beam will convert much of the bending moment in the beam into axial forces in the elastic support system and lead to significantly reduced deflections and bending moments for the beam.

Figure 6.1 shows two simply supported beams that have the same span of L and same rigidity of $E_b I$, subjected to the same uniformly distributed loads of q. Beam 2 is additionally supported at its centre by a vertical strut and two inclined strings. The strings have an elastic modulus of E_s and an area of A. To simplify the analysis while still capturing the physical essence of Beam 2, the axial deformation of the strut CD will not be considered. Beam 2 is called a beam-string structure in literature [6.1]. Calculate and compare the bending moments and deflections at the centre of the two beams.

Beam 1 (Figure 6.2a): The maximum bending moment and the maximum deflection occur at the centre of the beam and are respectively:

$$M_{1C} = \frac{qL^2}{8} \tag{6.1}$$

$$\Delta_{1C} = \frac{5qL^4}{384E_b I} \tag{6.2}$$

Beam 2 (Figure 6.2b): This is a statically indeterminate structure and the upward force from the strut needs to be determined before calculating the bending moment and deflection at the centre of the beam. The action of the strut CD on the beam can be replaced by a force, F_{CD}, that is to be determined. Figure 6.3 shows the geometrical relationship of string BD before and after deflection. The vertical deflection Δ_{2C} at the centre of the beam and the elongation δ of string BD has the following relationship:

$$\delta = \Delta_{2C} \sin\theta \tag{6.3}$$

The internal force in the string BD is:

$$F_{BD} = \frac{E_s A \delta}{L_{BD}} = \frac{E_s A}{L_{BD}} \Delta_{2C} \sin\theta \tag{6.4}$$

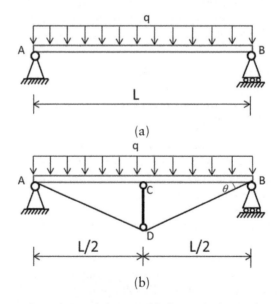

Figure 6.2 Two simply supported beams. (a) Beam 1: A simply supported beam. (b) Beam 2: A simply supported beam stiffened by a strut and two strings.

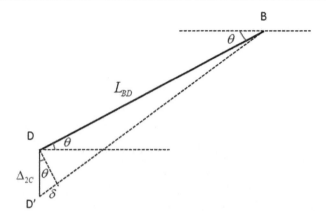

Figure 6.3 Geometrical relationship before (solid line) and after deflection (dashed line) of string BD.

The projection of the forces in the two symmetric strings to the vertical direction is equal to the force in the strut CD:

$$F_{CD} = 2F_{BD} \sin\theta = 2\frac{E_s A\delta}{L_{BD}}\sin\theta = \frac{2E_s A}{L_{BD}}\Delta_{2C} \sin^2\theta \qquad (6.5)$$

The deflection at the centre of the beam Δ_{2C} results from the actions of the downward uniformly distributed load, q, and the upward strut force, F_{CD}, generated from the two strings, $i.e.$

$$\Delta_{2C} = \frac{5qL^4}{384E_b I} - \frac{F_{CD}L^3}{48E_b I} \qquad (6.6)$$

Substituting F_{CD} in equation 6.5 into equation 6.6 gives:

$$\Delta_{2C} = \frac{5qL^4}{384E_b I} - \frac{L^3}{48E_b I}\frac{2E_s A_s}{L_{BD}}\Delta_{2C} \sin^2\theta \qquad (6.7)$$

Rearranging equation 6.7 and noting that $L/2 = L_{BD}\cos\theta$, gives the maximum deflection of Beam 2 as:

$$\Delta_{2C} = \frac{5qL^4}{384E_b I}\frac{1}{1+\dfrac{E_s AL^2 \sin^2\theta\cos\theta}{12E_b I}} = \Delta_{1C}\beta \qquad (6.8)$$

where:

$$\beta = \frac{1}{1+\dfrac{E_s AL^2 \sin^2\theta\cos\theta}{12E_b I}} = \frac{1}{1+\dfrac{2(E_s A/L_{BD})\sin^2\theta}{(48E_b I/L^3)}} = \frac{1}{1+\dfrac{2K_s \sin^2\theta}{K_b}} \qquad (6.9a)$$

and: $K_b = \dfrac{48E_b I}{L^3}$; $K_s = \dfrac{E_s A}{L_{BD}}$ (6.9b)

K_b is the flexural stiffness of the simply supported beam, K_s is the axial stiffness of string BD, and Δ_{1C} is the maximum deflection of Beam 1 defined in equation 6.2. Equation 6.8 indicates that the added strings and strut effectively reduce the maximum deflection of the original simply supported beam by a reduction factor, β. β in equation 6.9a is related to the ratio of the axial stiffness of the string to the flexural stiffness of the beam and the angle θ between the string and the beam. The term, $2K_s \sin^2 \theta$, in equation 6.9a can be interpreted as the spring stiffness in the vertical direction produced by the two inclined strings, K_{sv}. The reduction factor in equation 6.9a and the deflection in equation 6.8 can then be rewritten as:

$$\beta = \frac{1}{1 + \dfrac{2K_s \sin^2 \theta}{K_b}} = \frac{1}{1 + \dfrac{K_{sv}}{K_b}} = \frac{K_b}{K_b + K_{sv}}$$ (6.10a)

$$K_{sv} = 2K_s \sin^2 \theta = \frac{2E_s A}{L_{BD}} \sin^2 \theta$$ (6.10b)

$$\Delta_{2C} = \Delta_{1C} \frac{K_b}{K_b + K_{sv}}$$ (6.10.b)

When $K_{sv} = 0$, $i.e.$ no strut and strings, Beam 2 reduces to Beam 1. The deflection Δ_{2C} depends on the ratio of the flexural stiffness of the beam to the vertical stiffness of the strut and strings. For example, if $K_{sv} = K_b$, then $\beta = 1/2$ and $\Delta_{2C} = \Delta_{1C}/2$.

After introducing the vertical spring stiffness, $K_{sv} = 2K_s \sin^2 \theta$, of the strut and strings, Beam 2 in Figure 6.2b can be represented as a simply supported beam with a spring support at its centre as shown in in Figure 6.4. The strut and two strings effectively provide an internal support to the beam which can be converted to an external spring support to the beam to investigate the response of the beam.

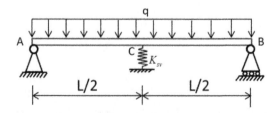

Figure 6.4 Alternative presentation of Beam 2 in Figure 6.2b.

F_{CD}, the spring force in equation 6.5, can be written as a function of Δ_{1C}

$$F_{CD} = \frac{2E_s A}{L_{BD}} \sin^2 \theta \cdot \Delta_{2C} = K_{sv} \Delta_{2C} = \Delta_{1C} \frac{K_{sv} K_b}{K_b + K_{sv}} \tag{6.11}$$

The bending moment at the centre of the beam-string structure is the summation of that induced by the uniformly distributed load and that induced by the concentrated force from the spring:

$$
\begin{aligned}
M_{2C} &= \frac{qL^2}{8} - \frac{F_{CD} L}{4} = \frac{qL^2}{8} - (\frac{5qL^4}{384E_b I} \frac{K_{sv} K_b}{K_b + K_{sv}}) \frac{L}{4} \\
&= \frac{qL^2}{8}(1 - \frac{5L^3}{4 \times 48E_b I} \frac{K_{sv} K_b}{K_b + K_{sv}}) = \frac{qL^2}{8}(1 - \frac{5}{4} \frac{K_{sv}}{K_b + K_{sv}})
\end{aligned}
\tag{6.12}
$$

When $K_{sv} = 0$, the beam-string structure reduces to the simply supported beam and $M_{2C} = M_{1C}$. When $K_{sv} = \infty$, the spring in Figure 6.3 becomes a roller support and the beam-string structure becomes a two-span beam and $M_{2C} = (qL^2/8)(1 - 5/4) = -qL^2/32$, which is just the bending moment at the fixed end of a propped cantilever with a span of $L/2$ subjected to a uniformly distributed load.

In the beam-string structure, the beam will also resist an axial compression force to balance the horizontal components of the string forces at the string-beam connections, which remove the need to create external supports to balance the string forces. Considering an arch as shown in Figure 6.5a, pin supports are required at the two ends of the arch to balance the horizontal forces generated by the arch, which tend to push outwards. If an arch-string structure, similar to the beam-string structure, is considered as shown

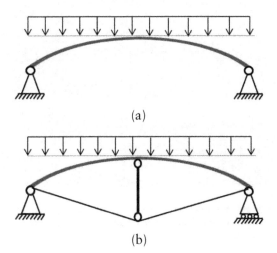

(a)

(b)

Figure 6.5 Two arches. (a) An arch structure. (b) An arch-string structure.

in Figure 6.5b, the horizontal components of the tension forces in the strings and compression forces in the arch will be balanced at their connection points. Thus, a roller support can be used at one of the two ends of the arch.

Figure 6.6 shows an application of adding profiled strings, in the forms of steel tendons, to an existing beam to increase the fundamental natural frequency of a floor system. The floor in a factory on which machines were operated on a daily basis experienced severe vibrations, causing significant discomfort for workers. It was found that resonance occurred when the machines operated. The solution was to avoid the resonance by increasing the stiffness of the floor and thus its fundamental natural frequency. It has been mentioned in Chapter 1 that reducing the maximum deflection of a structure means increasing its stiffness and hence the fundamental natural frequency of the structure.

Placing column supports would have reduced the span of the beam and produced a stiffer structure but this was not feasible due to the use of the area under the floor. Similar to the beam-string structure shown in Figure 6.2b, the externally added tendons provided two vertical elastic supports at the points where two steel bars (acting as struts) reacted against the concrete beams that support the floor. This produced a stiffer floor altering the fundamental natural frequency and solving the resonance problem.

Comparing the beam in Figure 6.1b and the real application shown in Figure 6.6, it can be noted that in the real application the inclined profiles of the tendons were created by the height of the beam rather than the height of

Figure 6.6 A floor beam is stiffened to form a beam-string structure to increase its fundamental natural frequency (Courtesy of Professor Jida Zhao, China Academy of Building Research, Beijing).

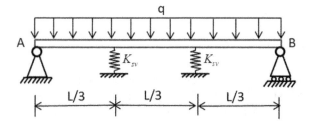

Figure 6.7 A simply supported beam with two equally spaced vertical elastic supports.

the strut, which creates two vertical elastic supports without taking the space under the beam. The equivalent model for the real application is shown in Figure 6.7 in which the tendons are converted to two vertical springs with stiffnesses of K_{sv} . Following equation 6.10b:

$$K_{sv} = \frac{2E_s A \sin^2 \theta}{\sqrt{(L/3)^2 + h^2}} = \frac{2E_s A}{\sqrt{(L/3)^2 + h^2}} \left(\frac{h}{\sqrt{(L/3)^2 + h^2}} \right)^2 \qquad (6.13)$$

Tendons were placed on each side of the beam (two tendons in all), and h is the height of the beam. The simply supported beam with two spring supports is a statically indeterminate structure. However, if the two spring forces of F can be determined, it becomes a statically determinate structure and the available equations for a simply supported beam can be used. The basic equations for calculating the deflection of a simply supported beam subjected to a uniformly distributed load q and to two symmetrically applied concentrated loads of F are respectively [6.2]:

$$v_q(x) = \frac{qx}{24E_b I}(L^3 - 2Lx^2 + x^3) \qquad (6.14)$$

$$v_F(x) = \frac{Fx}{6E_b I}[3Lx - 3x^2 + (L/3)^2] \quad L/3 \le x \le 2L/3 \qquad (6.15)$$

At $x = L/3$, the deflections due to the loads q and F are respectively:

$$v_q(L/3) = \frac{q(L/3)}{24E_b I}[L^3 - 2L(L/3)^2 + (L/3)^3] = \frac{11qL^4}{972E_b I} \qquad (6.16)$$

$$v_F(L/3) = \frac{F(L/3)}{6E_b I}[3L(L/3) - 3(L/3)^2 + (L/3)^2] = \frac{5FL^3}{162E_b I} \qquad (6.17)$$

The condition for the compatibility of deformations at $x = L/3$ is:

$$v_q(L/3) - v_F(L/3) = F / K_{sv} \tag{6.18}$$

This states that the difference between the deflections induced by q and F at $x = L/3$ is equal to the deflection of the elastic spring. Therefore, substituting equations 6.16 and 6.17 into equation 6.18 gives:

$$\frac{11qL^4}{972E_bI} - \frac{5FL^3}{162E_bI} = \frac{F}{K_{sv}} \tag{6.19}$$

The elastic spring force F can be determined from equation 6.19 as follows:

$$F = \frac{11qL^4}{972E_bI} \frac{K_{ba}K_{sv}}{K_{ba} + K_{sv}} = \frac{11qL}{30} \frac{K_{sv}}{K_{ba} + K_{sv}} \tag{6.20a}$$

$$K_{ba} = \frac{162EI}{5L^3} \tag{6.20b}$$

where K_{ba} is the point flexural stiffness of the beam or the inverse of the deflection at $x = L/3$ when the beam is subjected to two unit symmetric vertical forces at $x = L/3$ and $x = 2L/3$. When $K_{sv} = \infty$, it becomes a three, equal-span beam and each of the two middle roller supports will take 11/30 of the total loads. The spring force F depends on the ratio of the point flexural stiffness of the beam, K_{ba}, to the spring stiffness K_{sv}. The deflections at $x = L/3$ and at the centre of beam $(x = L/2)$ are:

$$\Delta_{L/3} = \frac{F}{K_{sv}} = \frac{11qL}{30} \frac{1}{K_{ba} + K_{sv}} \tag{6.21}$$

$$\Delta_{L/2} = \frac{5qL^4}{384E_bI} - \frac{23L^3}{432E_bI} \frac{11qL}{30} \frac{K_{sv}}{K_{ba} + K_{sv}}$$
$$= \frac{5qL^4}{384E_bI}(1 - \frac{1012}{675} \frac{K_{sv}}{K_{ba} + K_{sv}}) \tag{6.22}$$

To appreciate the effect of the internal vertical elastic supports, consider that the structure has the following estimated properties based on Figure 6.6: the span of the beam is $L = 6m$ and the cross-section of the beam is $b = h = 0.5\ m$, leading to a second moment of area of $I = 0.5 \times 0.5^3/12 = 5.208 \times 10^{-3}\ m^4$; the elastic modulus for the concrete beam and the elastic modulus for the tendons are respectively $E_b = 30 \times 10^9\ N/m^2$ and $E_s = 210 \times 10^9\ N/m^2$; the tendons have a diameter 20mm resulting in areas of $A = 314 \times 10^{-6}\ m^2$. The dead load including self-weight of the beam is $q = 100,000N\,/\,m$. Using equations 6.20b, 6.13, 6.20a, 6.14, 6.15, 6.21 and 6.22 derived previously:

$$K_{ba} = 2.344 \times 10^7 N\,/\,m; \quad K_{sv} = 3.763 \times 10^6 N\,/\,m; \quad F = 30436N;$$

$$v_q(L\,/\,2) = 10.8mm; \quad v_F(L\,/\,2) = -2.240mm;$$

$$\Delta_{L/3} = 8.09mm; \quad \Delta_{L/2} = v_q(L\,/\,2) - v_F(L\,/\,2) = 10.8 - 2.24 = 8.56mm$$

The fundamental natural frequencies before and after using the profiled tendons can be estimated using equation 1.9 as:

$$f_{bf} = 17.75\sqrt{\frac{1}{v_q(L\,/\,2)}} = 17.75\sqrt{\frac{1}{10.8}} = 5.40Hz$$

$$f_{af} = 17.75\sqrt{\frac{1}{\Delta(L\,/\,2)}} = 17.75\sqrt{\frac{1}{8.56}} = 6.07Hz$$

The ratio of the two natural frequencies is:

$$\frac{f_{af}}{f_{bf}} = \frac{6.07}{5.40} = 1.12$$

It can be noted that the use of the tendons increases the fundamental natural frequency by 12%, which was sufficient to solve the resonance problem [6.3].

6.2.2 Rigid Plates Supported by Vertical and Inclined Members

This example demonstrates the effectiveness and efficiency of inclined members compared to vertical members in resisting lateral deformation by converting bending moments to axial forces.

Figure 6.8 shows models of three plane structures in which a rigid plate is supported by four uniform members. The three structures have the same height, h, and all members have the same elastic modulus, E, and circular tubular section with a radius R and tube thickness t. They are subjected to the same lateral force P at the plate level. Model 1 is a typical frame structure, in which a rigid floor is supported by four vertical members that are rigidly connected to the plate and fixed to the ground. In Model 2, the rigid plate is supported by four members with pinned connections inclined at an angle of θ to the vertical. These members only experience tension or compression forces. Model 3 has the same geometry as that of Model 2, but with rigid connections. The lateral deflections of the three models will be compared.

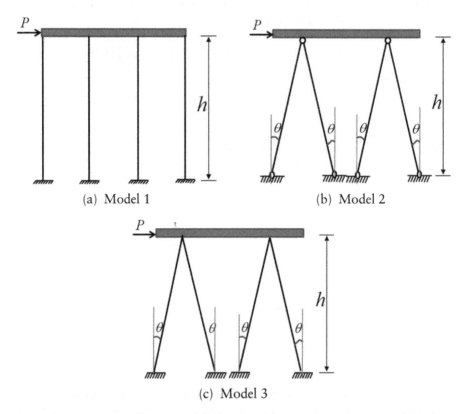

(a) Model 1   (b) Model 2

(c) Model 3

Figure 6.8 A rigid plate supported by four members. (a) Model 1: A rigid plate supported by four vertical members. (b) Model 2: A rigid plate supported by four inclined members with pin connections. (c) Model 3: A rigid plate supported by four inclined members with rigid connections.

The relative stiffness of the three models can be assessed qualitatively using the structural concept, *the more bending moments that are converted to axial forces, the smaller the deflection*. Model 1 (Figure 6.8a) will experience the largest lateral deflection of the three Models as the load *P* is transmitted to the ground through bending and shear in the four vertical members. Due to the pinned connections, Model 2 (Figure 6.8b) transmits the load *P* to the ground by the inclined members through tension and compression alone, which is far more effective than through bending and, as expected, Model 2 experiences smaller lateral deflection than Model 1. Model 3 transmits the load *P* through both axial forces and bending moments. The differences between Models 2 and 3 are the connections at the ends of the members. As Model 3 (Figure 6.8c) has stronger connections than Model 2, it is expected that Model 3 would experience less deflection than Model 2. Following this qualitative assessment, detailed analysis can be conducted to quantify the abilities of the three models to resist lateral deflection.

Model 1: A Rigid Plate Supported by Four Vertical Members
(Figure 6.8a)

As the two ends of the vertical members are rigidly connected with the plate
and the ground, the lateral stiffness of each of the four members is $12EI / h^3$.
Therefore, the lateral displacement of Model 1 due to load P is:

$$\Delta_1 = \frac{Ph^3}{48EI} \tag{6.23}$$

Model 2: A Rigid Plate Supported by Four Inclined Members with
Pin Connections (Figure 6.8b)

A rigid plate is supported by two identical pairs of inverted V shaped members.
Each pair of inclined members carry a half of the lateral load P, hence only the
two members of one of the inverted V shaped frames need be analysed. The
applied load and the internal forces in one inverted V shaped arrangement of
members are illustrated in Figure 6.9 in which the directions of the internal
forces are shown.

The internal forces in the two members, N_A and N_B, can be determined
from equilibrium conditions:

$$N_A \sin\theta + N_B \sin\theta = P / 2 \text{ and } -N_A \cos\theta + N_B \cos\theta = 0 \tag{6.24}$$

Solving the equations leads to:

$$N_A = N_B = \frac{P}{4\sin\theta} \tag{6.25}$$

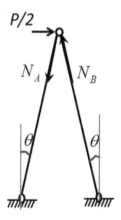

Figure 6.9 External and internal forces acting on the top node of the frame.

When $P/2$ is replaced by a unit force, the corresponding forces in the two members are:

$$\bar{N}_A = \bar{N}_B = \frac{1}{2\sin\theta} \tag{6.26}$$

The lateral deflection of Model 2 can be calculated using Equation 2.14 as follows:

$$\Delta_2 = \sum \frac{N_i \bar{N}_i L}{EA} = \frac{1}{EA} \frac{P}{4\sin\theta} \cdot \frac{1}{2\sin\theta} \cdot \frac{h}{\cos\theta} \times 2 = \frac{Ph}{4EA} \frac{1}{\sin^2\theta\cos\theta} \tag{6.27}$$

Model 3: A Rigid Plate Supported by Four Inclined Members with Rigid Connections (Figure 6.8c)

Model 3 looks like Model 2, except the connections to the plate and the ground are rigid. The equilibrium equation in the lateral direction for Model 3 can be found from the element stiffness matrix of an inclined beam member in finite element analysis [6.4] as follows, when no rotation and axial deformation are considered at the top node of the members:

$$4(\frac{EA}{L}\sin^2\theta + \frac{12EI}{L^3}\cos^2\theta)\Delta_3 = P \tag{6.28}$$

where L is the length of the inclined member and can be expressed as $h/\cos\theta$; the terms in the bracket represent the lateral stiffness of a single inclined uniform beam. Solving this equation gives:

$$\Delta_3 = \frac{P}{4(\frac{EA}{L}\sin^2\theta + \frac{12EI}{L^3}\cos^2\theta)} = \frac{P}{4(\frac{EA}{h}\sin^2\theta\cos\theta + \frac{12EI}{h^3}\cos^5\theta)} \tag{6.29}$$

When the rigid connections reduce to pinned connections, i.e. the members are not be able to transmit bending, $I = 0$, and equation 6.29 reduces to equation 6.27 for Model 2. When $\theta = 0°$, the inclined members become vertical members and equation 6.29 reduces to equation 6.23 for Model 1. Comparing equations 6.27 and 6.29, it can be seen that when $0 \le \theta < 90^0$, $\Delta_3 < \Delta_2$, i.e.:

$$\Delta_3 = \frac{P}{4(\frac{EA}{h}\sin^2\theta\cos\theta + \frac{12EI}{h^3}\cos^5\theta)} = \frac{Ph}{4EA\sin^2\theta\cos\theta} \frac{1}{(1 + \frac{3I}{Ah^2}\text{ctan}^2\theta\cos^2\theta)}$$

$$= \Delta_2 \frac{1}{(1 + \frac{3I}{Ah^2}\text{ctan}^2\theta\cos^2\theta)} < \Delta_2 \tag{6.30}$$

The lateral deflections of the three Models are shown in equations 6.23, 6.27 and 6.29. The ratios of the deflections of the three Models can be obtained by substituting $I = \pi R^3 t$ and $A = 2\pi R t$ for a circular tube section when $R \gg t$, as:

$$\frac{\Delta_2}{\Delta_1} = \frac{Ph}{4EA} \frac{1}{\sin^2\theta\cos\theta} \cdot \frac{48EI}{Ph^3} = \frac{12I}{Ah^2} \frac{1}{\sin^2\theta\cos\theta} = \frac{6R^2}{h^2} \frac{1}{\sin^2\theta\cos\theta} \quad (6.31)$$

$$\frac{\Delta_3}{\Delta_1} = \frac{6R^2}{h^2} \frac{1}{\sin^2\theta\cos\theta} \frac{1}{(1 + \dfrac{3R^2}{2h^2}\mathrm{ctan}^2\theta\cos^2\theta)} \quad (6.32)$$

$$\frac{\Delta_3}{\Delta_2} = \frac{1}{(1 + \dfrac{3R^2}{2h^2}\mathrm{ctan}^2\theta\cos^2\theta)} \quad (6.33)$$

Consider the cases of $R = 100mm$, $h = 4000mm$ and $8000mm$ and $\theta = 5^0$, $10^0, 15^0, 30^0$ and 45^0, and the deflection ratios based on equations 6.31–6.33 are listed in Table 6.1.

It can be noted from Table 6.1 for this particular case that:

1. Converting bending moments to axial forces makes the structures much stiffer and significantly reduces the deflections due to the lateral loading. For the case of $h = 4000mm$ and $\theta = 15^0$, the reduction is approximately 94%.
2. Even a small inclination angle (5^0 from the vertical) can still reduce the lateral deflection by over 50% for $h = 4000mm$ and by over 87% for $h = 8000mm$.
3. When the inclination angle is larger than or equal to 10^0, the effect of the rigid connections of the inclined members is negligible for reducing the lateral deflection.
4. When the structure becomes higher, the inclined members become even more effective and efficient to resist lateral deflection than normal vertical

Table 6.1 The Deflection Ratios for Different Angles of Member Inclination and for Two Different Heights of Structure

$h = 4000mm$	$\theta = 5^0$	$\theta = 10^0$	$\theta = 15^0$	$\theta = 30^0$	$\theta = 45^0$
Δ_2/Δ_1	0.4956	0.1263	0.05796	0.01732	0.01061
Δ_3/Δ_1	0.4419	0.1226	0.05726	0.01728	0.01060
Δ_3/Δ_2	0.8916	0.9716	0.9880	0.9979	0.9995
$h = 8000mm$	$\theta = 5^0$	$\theta = 10^0$	$\theta = 15^0$	$\theta = 30^0$	$\theta = 45^0$
Δ_2/Δ_1	0.1239	0.0315	0.01448	0.004330	0.002652
Δ_3/Δ_1	0.1202	0.0313	0.01444	0.004328	0.002651
Δ_3/Δ_2	0.9705	0.9927	0.9970	0.9994	0.9999

columns. For $h = 8000mm$ and $\theta = 15^0$, the deflection ratio of Model 2 to Model 1 is only 0.014 and there is little difference between rigid and pinned connections for Models 2 and 3.

This quantitative analysis of the three models provides a theoretical basis to explain that inclined members can be effectively used, replacing vertical columns, to support upper structures against lateral deflection. Therefore, it can be concluded that it is very effective and efficient, where possible, to use inclined bar members to replace convention vertical columns in which the fourth concept is embedded.

6.3 Practical Examples

6.3.1 Structures with Vertical Internal Elastic Supports

6.3.1.1 Spinningfields Footbridge, Manchester

The Spinningfields Footbridge over the River Irwell, linking Spinningfields in Manchester and New Bailey in Salford, UK, has a single span of 44 meters and was built in 2012. It can be seen from Figure 6.10a that it has the appearance of a light and elegant steel footbridge. The footbridge consists of a bridge deck, a group of beams, a series of struts and a cable or a tendon. The struts provide links between the cable and the beams that support the deck (Figure 6.10b), and the different heights of the struts create the profile of the bridge.

The bridge experiences bending with the cable carrying a tension force while the beam with a circular section beneath the deck carries a compression force to balance the bending moment induced by vertical loads. The distance between the cable and the circular section beam is largest at the centre of the bridge and gradually reduces toward the bridge supports, which reflects the profile of the bending moment diagram for a simply supported beam subjected to uniformly distributed loads.

The circular section beam, which directly supports the deck, is supported by a series of struts that are equivalent to vertical internal elastic supports, which effectively reduce the bending moments and deflections of the beam.

To gain a better understanding of the behaviour of the footbridge, consider two simplified cases which still capture the physical essence of the footbridge, as shown in Figure 6.11. Model 1 is a normal, simply supported beam subjected to uniformly distributed load, and Model 2 is a beam-string structure. The basic data used for analysis are the span of $L = 40m$, the elastic modulus of $E = 200 \times 10^9 N/m^2$ and the distributed load of $q = 10kN/m$. The cable has a parabolic shape with sag of 2m. Assume the beam has a solid section of $b \times h = 800mm \times 400mm$ and the struts and the cable have solid circular sections with diameters of 80mm and 40mm respectively. Seven struts are uniformly distributed along the length of the beam at an interval of 5m and follow a parabola with a sag of 2m.

(a)

(b)

Figure 6.10 The Spinningfields Footbridge, Manchester. (a) Overall view. (b) Detailed view shown the relationship between the cable, struts, beams and the deck.

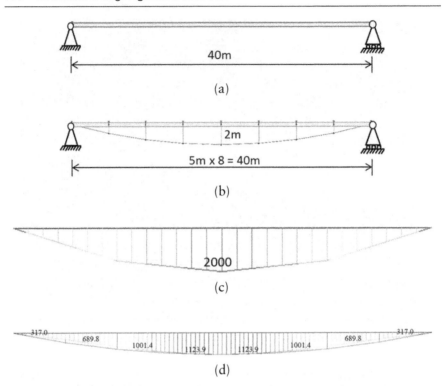

40m

(a)

2m

5m x 8 = 40m

(b)

2000

(c)

317.0   689.8   1001.4   1123.9   1123.9   1001.4   689.8   317.0

(d)

Figure 6.11 Effectiveness and efficiency of a beam-string structure. (a) Model 1: A simply supported beam. (b) Model 2: A beam-string structure. (c) Bending moment diagram for Model 1. (d) Bending moment diagram for Model 2.

Figure 6.11c and 6.11d show the bending moment diagrams of the simply supported beam and the beam-string system with the maximum values of 2000kNm and 1139kNm respectively. The corresponding maximum deflections of the two models are 0.391m and 0.193m respectively. These results indicate that a beam-string system can be designed much lighter than a corresponding beam system.

Beam-string structures are often used for roof structures. Figures 6.12a and 6.12b show that the roof structure of the Shanghai Pudong Airport Terminal 1 consists of a series of beam-string beams in which the strings and the struts can be easily identified. Figure 6.12b also shows several inclined cables anchored on a column which provide the structural stiffness in the two horizontal directions of the roof and increase the vertical resistance of the roof to wind uplift.

6.3.1.2 The Roof of the Badminton Arena for the 2008 Olympic Games, Beijing

The beam-string structures of the footbridge and the terminal roof described in Section 6.3.1.1 are plane, 2D, structures. However plane beam-string structures

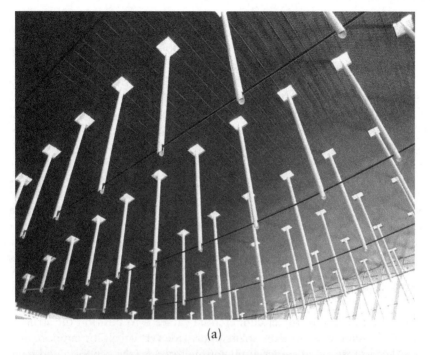

(a)

(b)

Figure 6.12 The roof at the Shanghai Pudong Airport Terminal 1. (a) A series of beam-string beams. (b) Inclined cables that increase the lateral stiffness of the roof and provide vertical resistance to uplift wind loading.

have been developed into three dimensional structures forming so called string supported spherical shells [6.5] or cable supported domes [6.6]. The roof of the Badminton Arena (Figure 6.13) for the 2008 Olympic Games in Beijing is a string supported dome that spans 98m.

To understand the structural components and behaviour of the Arena roof, a similar but simpler example is illustrated in Figure 6.14. The roof consists of a single-layer shell with struts and cables in both circumferential and radial directions. The top ends of the struts are connected to the shell and their lower ends are linked with both radial and circumferential cables. Figure 6.14a shows that the roof has three layers of circumferential cable rings.

The cross-section of the string supported shell roof (Figure 6.14a) looks like an arch supported by struts at three different levels. The load paths or internal force paths of the roof structure are direct and clear. Most of the external loads applied on the shell are transmitted to the struts and through the struts to the cables. At the highest level the action of radial and circumferential cables in space act as a series of plane beam-string structures (Figure 6.12a). The forces from the two struts are balanced at their connection points to the radial and circumferential cables and transmitted by radial cables to the struts in the next lower level. This type of force transmission continues to the lowest level of struts. The function of the circumferential cables is to position the struts and the radial cables and allow the struts to provide vertical elastic supports to the shell. The lowest radial cables apply tensile forces to the supports and tend to

Figure 6.13 The roof of the Badminton Arena for the 2008 Olympic Games, Beijing.

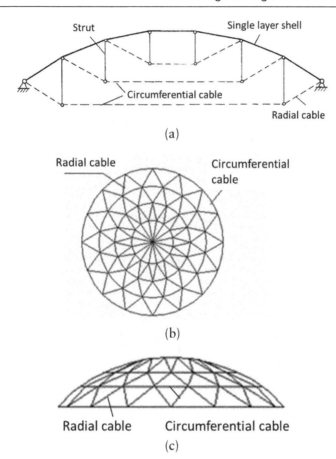

Figure 6.14 A string supported shell roof. (a) Cross-section. (b) Plan. c) Elevation
[6.5] (Courtesy of Professor Zhihua Chen, Tianjin University, China).

pull the supports inward while the shell acts in compression tending to push
the supports outward. Therefore, the two sets of forces are partly self-balanced
and produce smaller reactions on the ring beam. As the struts locate on the
radial and circumferential cables, they act as internal vertical elastic supports
to the shell which leads to smaller internal forces in the shell and hence smaller
deflections. The performance of string supported shell roofs can be further
improved by applying pre-stressing to structural members to produce internal
forces in the members to counteract those induced by external loads which
could lead to even more efficient structures.

The analysis of a string supported shell roof needs the use of a computer but
much of the structural behaviour of the shell roof can be illustrated using the
example of the beam-string structures examined in Section 6.2.1.

Back to the Badminton Arena for the 2008 Olympic Games, Figures 6.15a
and 6.15b show the plan and the cross-section of the roof structure. As
shown in Figure 6.15b, the Badminton Arena has five rings of circumferential

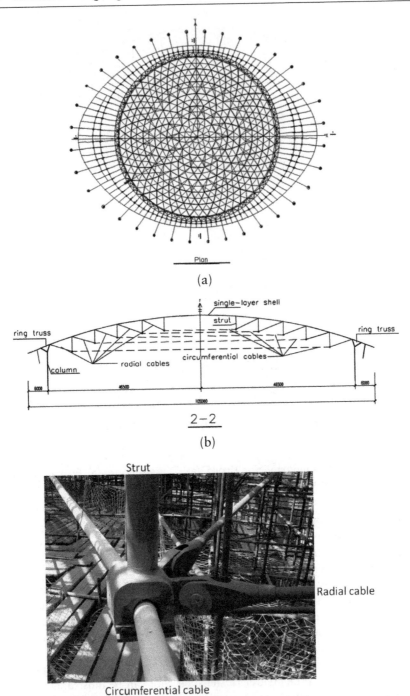

Figure 6.15 (a) Plan of the Arena roof. (b) Cross-section of the Arena roof. (c) A connection between a circumferential cable, two radial cables and a strut
(Courtesy of Professor Ailin Zhang, Beijing University of Technology, China).

cables at different levels under the single-layer shell, which are linked to the shell by radial cables and struts. In construction, stressing the circumferential cables produced tensile forces in the radial cables and compression forces in the struts supporting the upper single-layer shell. To make the construction process more convenient, the circumferential cables were stressed at four tensioning points provided in each cable to reduce the connection friction between the struts and cables. Figure 6.15c shows a typical connection between a strut, a circumferential cable and two radial cables, indicating the internal force paths.

As the struts positioned and supported by radial and circumferential cables at five different levels, provide many vertical internal supports to the roof dome, the roof is able to cover a huge area without using any inner supports.

6.3.2 Structures Supported by Inclined Members

6.3.2.1 Three Types of Support to Superstructures

Figure 6.16a shows the external view of an airport terminal. The roof is supported by slender, inclined members, which have pin connections at their two ends and thus carry only axial forces. Roofs and floors are normally supported by vertical columns and horizontal beams, which form frame structures to transmit both vertical and lateral loads to their supports. For example, Figure 6.16b shows a building structure in which columns are the main load bearing members with the externally exposed columns supporting the upper storeys of the building. The two ends of the columns can be considered to have rigid connections. The cross-section size of the columns and the distance between adjacent columns leads to a sense of solidity of the building. A combination of the supporting systems in Figures 16a and 16b (*i.e.* pin ended inclined members and rigidly connected vertical members) leads to inclined members with rigid connections. Figure 16c shows a building in which the upper structure is supported by a series of V or inverted V shaped columns in which their bottom and top ends are close to rigid connections.

(a)

Figure 6.16 Comparison of three supporting systems. (a) The roof of an airport terminal supported by inclined pin ended members. (b) Vertical column members support the upper storeys of a building. (c) Inclined members support the upper storeys of a building.

(b)

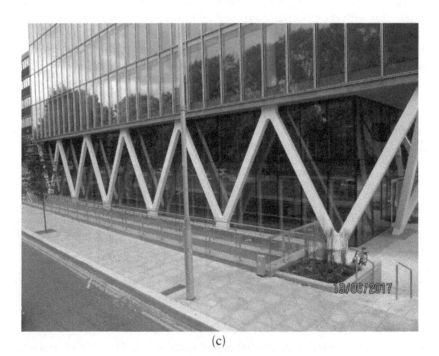

(c)

Figure 6.16 (Continued)

The three models (Figure 6.8) qualitatively and quantitatively studied in Section 6.2.2 are abstracted from the three structures in Figure 6.16, which have demonstrated that the use of inclined members is very effective and efficient for resisting lateral loading. The terminal roof (Figure 6.16a) has a height of over 20m, and the results in Table 6.1 have shown that inclined members become even more effective and efficient as the height increases.

6.3.2.2 Ontario College of Art and Design, Toronto

Figure 6.17 shows the Sharp Centre for Design, an expansion of the Ontario College of Art and Design (OCAD) in Toronto. It looks like a huge rectangular block that is in fact a two-storey building 80m long and 30m wide. When viewed from a distance, the block appears similar to a tabletop floating over the ground because it is only supported by a small number of slender members and cantilevered concrete walls, which support stairs between the ground and the block.

The block is actually supported by twelve 29m long steel legs each with a diameter of 914mm and a wall thickness of 25mm, which seem to be randomly arranged [6.7]. The 12 legs form six pairs of triangular leg arrangements to achieve better stability and lateral resistance. The steel columns are

Figure 6.17 The Sharp Centre for Design, Toronto (Courtesy by Mr. Nicolas Janberg, structurae.net, Germany).

tapered at the upper and lower ends indicating that they act as pinned connections to the ground and to the block and act as compression members rather than bending members. The planes of the two front pairs of legs orient ± 45 degrees from the central axis in the longitudinal (longer) direction and are perpendicular to each other, which provide the lateral stiffness in the two horizontal directions, in addition to their vertical supports to the block. The two middle pairs of triangular legs are only orientated in the transverse (shorter) direction, providing lateral stiffness in this direction. Due to the asymmetric position of the concrete stair-core to the block, the two other pairs of legs are arranged opposite to the concrete core, along the longitudinal direction and leaning inward in the transverse direction to provide lateral stiffness in both transverse and longitudinal directions. These two pairs of legs also compensate for the effect of the asymmetric location of the large, stiff, concrete stair-core that also contributes lateral stiffness in both transverse and longitudinal directions.

At first sight, it is puzzling how twelve slender inclined (in the longitudinal direction) members can safely support the large structure of the Sharp Centre for Design. However, the results presented in Table 6.1 effectively explain the technical feasibility of using slender inclined, pin ended members to replace more conventional columns.

6.3.2.3 Roof Supports of Terminal 5 at Heathrow Airport, London

Another good example of using long inclined pin ended members to replace vertical columns can be seen in the Heathrow Terminal 5 as shown in Figure 6.18a. A series of pairs of long inclined members are placed along a window wall to support the upper structure systems, including long span beams and the roof of the terminal. Examining one typical unit of six connected inclined members, Figure 6.18a shows a pair of long steel tubes forming a triangular shape are placed next to the window wall along the longitudinal (longer) direction and inclined inward in the transverse direction. The bottom ends of the two tubes are pinned to their foundations while their top ends are pin connected together with two upper pairs of inclined steel tube members (Figure 6.18a). The top ends of the shorter pair of inclined members on the top right (Figure 6.18a) provide the supports at the ends of two adjacent roof beams and are linked by a horizontal bar member (Figure 6.18b). The longer pair of inclined tubes acts as internal props to the two roof beams and provide the end supports to the cables that are used to stiffen the roof beams. A horizontal bar connects the top ends of the two inclined props which not only positions the props in the longitudinal direction but also provides lateral supports to the two roof beams.

The three pairs of inclined members, two horizontal members and two roof beams form a stable, equilibrium and mutual-supporting system. The long steel members are used to support the two upper pairs of the inclined members that in turn support the two roof beams. It is also true that the top ends of the long steel members are positioned and supported by the lower ends of the two pairs of inclined upper members, while the top ends of these inclined members are positioned and supported by the two roof beams and the two horizontal bars.

(a)

(b)

Figure 6.18 Inclined pin ended members supporting the long span roof at Terminal 5 Heathrow Airport. (a) A series of pairs of pin ended inclined steel tubes with two pairs of upper inclined members supporting the roof structure. (b) The two upper pairs of inclined members provide four supports to two adjacent roof beams.

Considering the large dimensions of the terminal building (Figure 6.18), the long inclined support members appear slender and sparsely spaced, which is due to the effectiveness and efficiency of the inclined members that are subjected to axial forces rather than bending action demonstrated in Section 6.2.2 and due to the self-supporting of the two roof beams, six incline members and the two horizontal members.

6.3.3 Using Self-Weight of Structural Members—Alamillo Bridge, Seville

The Alamillo Bridge is a well-known example in which the self-weight of the pylon is used to balance the self-weight of the bridge deck and part of the live loads on the bridge. The Alamillo Bridge, shown in Figure 6.19, is a cable-stayed bridge that was one of the six bridges built to improve infrastructure for the Expo 1992 on the island of La Cartuja, just outside the city of Seville [6.8, 6.9]. The bridge has a span of 200m and is supported at the two ends and by 13 pairs of cables with a uniform spacing of 12m.

The original idea of the design came from Santiago Calatrava arranging for the forces in the cables supporting the bridge deck to be balanced by the considerable self-weight of a massive reinforced concrete pylon with a backward inclination of 58 degrees from the ground, rather than the traditionally used back-stay cables [6.8]. This idea of the design is illustrated in Figure 6.20.

Normally, a pylon for a cable stayed bridge is a vertical member and cables are arranged at both sides of the pylon that transfers the compression forces

Figure 6.19 The Alamillo Bridge, Spain (Courtesy of Mr Per Waahlin, Sweden).

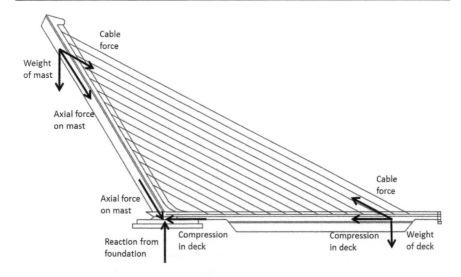

Figure 6.20 Illustration of the forces acting on the Alamillo Bridge (this figure is produced based on Figure 2 in [6.9]).

and bending moments generated from the cables to its foundation. The lateral force components generated by the two sets of cables on the pylon are in opposite directions and partly self-balancing, leading to smaller bending moments in the pylon. The unique design of the inclined pylon (Figure 6.20) has two effects on internal force paths:

1. Globally, the weight of the pylon would be designed to balance part of the loads from the bridge deck, self-weight and live loads.
2. Locally, the resultant forces of the self-weight of the pylon and the cable forces resulting from the bridge deck loads pass through the central axis of the inclined pylon, leading to that the pylon experiences mainly compressive forces rather than bending moments. In other words, the bending moments in the pylon induced by the cable forces are partly balanced by the bending moments generated by the self-weight of the pylon, thus converting bending moments to compressive forces in the pylon.

To achieve this fully would require a pylon of extreme size and mass to balance the deck forces and create the ideal compression only force in the pylon. Theoretically, it might be possible to achieve self-balancing and an ideal state of internal forces for one particular loading case. However, for a bridge subjected to a wide range of loading scenarios, which need to be considered in the design of civil engineering structures, some bending moments would always need to be allowed in the design of the pylon. The idea of an inclined pylon nevertheless does help to reduce the bending moments in the pylon.

6.4 Further Comments

The Y shaped column subjected to vertical loads and inclined members subjected to lateral loads are examined quantitatively and independently in Sections 4.2.2 and 6.2.2. The understanding gained from these two sections can be used to judge the behaviour of existing structures. Large V or Y shaped supports can be seen at 1 Spinningfields, Manchester, as shown in Figure 6.21. The cross-section of the inclined members gradually becomes smaller with an increase in height, which leaves an impression that the inclined members experience their smallest bending moments at their upper ends and the largest bending moments at their lower ends, *i.e.* the variable cross-section of the members seems to suggest that they are subjected to bending. The top ends of the members support and link to floor beams that in turn restrain the relative lateral deflection between the two top ends of the members for vertical loading. The example of the tied Y column in Section 4.2.2 indicates that the arms are subjected mainly to compression forces rather than bending moments due to vertical loading. Considering the actions of lateral loads, the Y columns are similar to that in Figure 6.8c, when the floor supported by the Y columns is considered as a rigid plate. Table 6.1 shows that at an inclination angle of the members of 45 degrees to the vertical there would be little bending moments in the members, and thus, from a structural point of view alone, the arms of the Y shaped columns could be designed with a constant cross-section.

Figure 6.21 Variable cross-section of the arms of Y shaped columns in a building, UK.

Figure 6.22 The roof supports of the Xi'an North Railway Station, China. (a) Overview of the roof and its supports. (b) A typical Y branch support and overhangs.

The physical measures mentioned in the earlier chapters can be jointly used to achieve more efficient structures. Figure 6.22 shows the waiting hall of the Xi'an North Railway Station, China. The roof of the hall is a large-span light steel folding plate grid structure that is hidden by the suspended ceiling, but the supports to the roof can be seen. Figure 6.22b gives a close look at one of the supports to the roof structure. Four inclined members, pin connected at the top of a column, provide four external pin supports to the roof grid structure, allowing a larger distance between columns.

The physical measures used in the roof structure include overhangs to reduce the span between supports, the Y shaped columns for providing more supports to the roof and inclined bar members.

References

6.1   Saitoh, M. and Okada, A. The Role of String in Hybrid String Structure, *Engineering Structures*, 21, 756–769, 1999.

6.2   Gere, J. M. and Timoshenko, S. P. *Mechanics of Materials*, PWS-KENT Publishing Company, USA, 1990.

6.3   Ji, T., Bell, A. J. and Ellis, B. R. *Understanding and Using Structural Concepts*, Second Edition, Taylor & Francis, USA, 2016.

6.4   Cook, R. D., Malkus, D. S. and Plesha, M. E. *Concepts and Applications of Finite Element Analysis*, John Wiley & Sons, USA, 1989.

6.5   Chen, Z. *Cable Supported Domes*, Science Press, Beijing, China, 2010.

6.6   Zhang, A. Olympic Badminton Area: Cable Suspended Dome, *The Structural Engineer*, 85(22), 23–24, 2017.

6.7   Silver, P., Mclean, W. and Evans, P. *Structural Engineering for Architects: A Handbook*, Laurence King Publishing Ltd., London, 2013.

6.8   Aparicio, A. C. and Casas, J. R. The Alamillo Cable-Stayed Bridge: Special Issues Faced in the Analysis and Construction, *Structures and Buildings, the Proceedings of Civil Engineers*, 122, 432–450, 1998.

6.9   Guest, J. K., Draper, P. and Billington, D. P. Santiago Calatrava's Alamillo Bridge and the Idea of the Structural Engineer as Artist, *ASCE, Journal of Bridge Engineering*, 18(10), 936–954, 2013.

Chapter 7

Concluding Remarks

The contents of Chapters 2 to 6 can be summarised in the hierarchical relationships shown in Figure 7.1 in which some of the practical examples in these chapters are listed.

Figure 7.1 shows paths and connections from theory (the Principle of Virtual Work) to four structural concepts, then on to several routes to implementation and finally to a large number of practical application cases indicating that the four structural concepts have a range of applications for structural design against deflection. It is noted that this observation is applicable to other structural concepts, as it is natural that good structural concepts can lead to wide and wise applications.

Figure 7.1 also illustrates one type of relationship between theory and practice, *i.e.* moving from theory downward to practice applications. However, the arrows in Figure 7.1 can also be reversed and presented in another type of relationship, *i.e.* moving from practical cases upward to theory. In other words, implementation measures can be identified from practical cases and structural concepts can then be abstracted from the implementation measures. The downward and upward relationships between theory and practice are complementary to each other and can enrich both practical applications and theoretical studies. The presentation in this book takes the downward approach from theory to practice, and the implementation measures are developed based on the structural concepts.

Some implementation measures, however, were created intuitively to solve problems encountered in practice. For example, to reduce the horizontal thrusts on the foundations from the two inclined arches of the Rayleigh Arena discussed in section 5.3.2.2, (Figure 5.13), tendons were provided between the ends of the two arches to balance part of the thrusts and thus the foundations experienced much smaller horizontal forces from the arches. Further study of the action of the tendons also led to the implementation routes of self-balancing of internal forces and the provision of internal elastic supports and then to the structural concept of smaller internal forces leading to smaller deflections.

The four structural concepts provide a basis for creative structural design against deflections for tall buildings, long-span bridge/roofs and other structures that are sensitive to deflections. Some implementation routes and physical measures have been explicitly explained and provided through illustrated

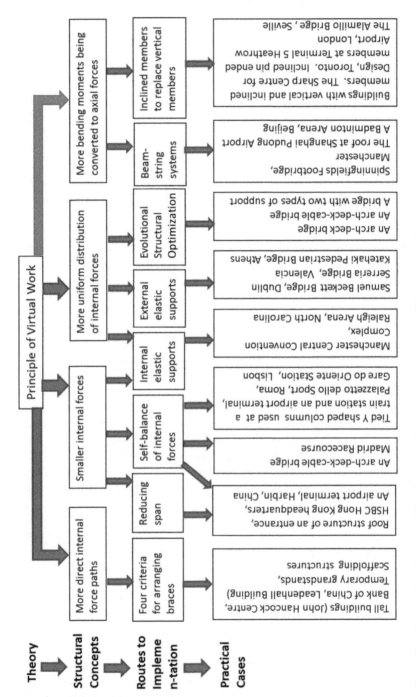

Figure 7.1 Hierarchical relationships between theory and practice and between book contents.

hand calculations and practical examples. According to Figure 7.1, there are four structural concepts, more routes to implementation and even more physical measures implemented into practical cases, which is not an exclusive list. In other words, there are opportunities to generate new routes to implementation, create new implementation measures or use existing ones for achieving smaller deflections of structures and for designing more effective and efficient structures that are likely to appear more elegantly.

Bibliography

Allan, E. *et al. Form and Forces: Designing Efficient, Expressive Structures*, John Wiley & Sons, USA, 2009.

Balmond, C. *Informal*, Prestel, Germany, 2002.

Charleson, A. *Structure as Architecture—A Source Book for Architects and Structural Engineers*, Architectural Press, UK, 2005.

Frei, O. and Bodo, R. *Finding Form: Towards an Architecture of the Minimal, Deutscher Werkbund Bayern*, Edition Axel Menges, Third Edition, 1996.

Heyman, J. *Structural Analysis: A Historical Approach*, Cambridge University Press, UK, 1998.

Jennings, A. *Structures—From Theory to Practice*, Spon Press, London, 2004.

Macdonald, A. J. *Structure & Architecture*, Second Edition, Architecture Press, Oxford, 2003.

Margolius, I. *Architects + Engineers = Structures*, Wiley-Academy, UK, 2002.

Parkyn, N. *The Seventy Architectural Wonders of Our World*, Thames & Hudson, London, 2002.

Rappaport, N. *Support and Resist: Structural Engineers and Design Innovation*, The Monacelli Press, USA, 2007.

Robinson, D. N. *Consciousness and Its Implications*, The Teaching Company, USA, 2007.

Rosenthal, H. W. *Structural Decision*, Chapman & Hall Ltd., London, 1962.

Salvadori, M. and Heller, R. *Structures in Architecture: The Building of Buildings*, Prentice-Hall, NJ, USA, 1986.

Sandarker, B. N. *On Span and Space: Exploring Structures in Architecture*, Routledge, London, 2008.

Sandaker, B. N., Eggen, A. P. and Cruvellier, M. R. *The Structural Basis of Architecture*, Second Edition, Routledge, London, 2011.

Schlaich, J. and Bergermann, R. *Light Structures*, Prestel, Germany, 2004.

Schlaich, M. Elegant Structures, *The Structural Engineer*, 2015.

Silver, P., Mclean, W. and Wvans, P. *Structural Engineering for Architects: A Handbook*, Laurence King Press, London, 2013.

Sprott, J. C. *Physics Demonstrations—A Sourcebook for Teachers Physics*, The University of Wisconsin Press, USA, 2006.

Uffelen, C. V. *Bridge: Architecture + Design*, Braun Publishing AG, Switzerland, 2009.

Young, J. W. *A Technique for Producing Ideas*, McGraw-Hill, USA, 2003.

Index

Page numbers in italics indicate figures; page numbers in bold indicate tables.